About this book

This book is the third in a series of three books covering Years 7–10 aimed at the top 10% or so of pupils aged 11–15. The series has been written and produced so that successive volumes provide an enrichment programme for the most able pupils in the lower secondary school, though many items could be adapted for use with other age groups and with a wider ability range. Where pupils are working mainly on their own, they will get more out of what Book 3 has to offer if they have first tackled a good part of the material in Books 1 and 2. Each individual section in this book can also be used to form the basis of an excellent 'masterclass' for secondary pupils.

Each book in the series contains twenty or more sections together with Comments and Solutions. These Comments and Solutions provide more than just a convenient way for pupils to check their answers: they have been written to help provoke further thought and discussion with peers or with a teacher. Teachers may have to stress that, even when pupils think they have succeeded in solving a problem, it is important for them to reflect on the approach they have used (and the reason why it works). In particular, they should never be satisfied with merely getting the answer.

The material has been extensively trialled and revised in the light of teachers' comments. The styles of the different sections are very varied. This provides a freshness of approach, which not only adds to the value of the collection, but which challenges pupils to think about familiar topics in new ways. Though different items are written in different styles, all sections seek to challenge able pupils in ways that will help them to:

• develop a deeper understanding of the ideas and methods which are central to secondary mathematics

• lay a firmer foundation for later work

• develop reasoning skills

• cultivate a broader interest in mathematics

• develop their ability to research and to interpret mathematics.

The potential of the material is illustrated by three comments from teachers who trialled one batch of draft material.

'Excellent item on a topic that is rarely understood in depth.'

'If only my A-level students had seen this in Year 7!'

'This item is nicely constructed in a way which makes it suitable for either individuals or small groups working on their own.'

Other materials and strategies which are currently used to extend more able pupils tend either to provide interesting 'time-fillers' which have little to do with ordinary curriculum work, or to accelerate pupils through the National Curriculum levels (and the associated examinations) ahead of their peers. In contrast, this book contains activities which encourage pupils to think more deeply about relatively elementary mainstream material while covering a fairly broad range of standard curriculum topics. The book also includes some extension material which enriches the standard curriculum. Thus it offers schools an alternative both to using time-fillers, and to the acceleration strategy referred to above (a strategy which puts considerable burdens on staff, and which leads to able pupils being permanently out of phase with their peers).

Contents

How to use this book

To teachers and parents

- This book is aimed at pupils with the same kind of background as the top 10% of pupils in Years 9–10 in English schools (i.e. pupils aged 13–15). It is the third of a series of three books containing material to challenge and extend such pupils in Years 7–10 (aged 11–15).

- To get the most out of what Book 3 has to offer, pupils should first tackle a good part of Books 1 and 2.

- Many of the sections included can be completed by pupils working on their own; the **Comments and Solutions** provide additional support. However, all sections would benefit from some discussion – either with other pupils as part of the activity, or with a teacher or parent.

- A brief summary of the intended learning objectives is given at the beginning of each section, along with the intended classroom organisation (whether individual, or small group work), any necessary materials which must be available, and any sections which should have been tackled before you start.

  This information is provided both to help the experienced teacher, who may be using the material for the first time, and to alert pupils to any necessary prerequisites.

- When pupils do not understand a technical term which appears in the text, they should first consult the **Glossary** at the back of the book.

- The amount of time a given pupil spends on each section will vary. However, we anticipate that each shorter section should take 45–60 minutes, while each longer section may take 1–2 hours.

- Some sections explicitly tell pupils to *choose* a task which interests them, rather than to attempt every single task. Even where pupils are expected to work systematically through a section, it is not essential for them to complete every single question.

How to use this book

To pupils

- These activities challenge you to think more seriously about some of the mathematics you have learned, and about the way you do mathematics. This will not only improve what you can do this year, but will help to lay a solid foundation for future work.

- To get the most out of what Book 3 has to offer, you should first tackle a good part of Books 1 and 2.

- Some problems are straightforward; other problems are rather hard. Don't give up too easily. If the last couple of questions in a section are too difficult, or involve ideas which you have not yet met, don't worry. It is more important to think carefully about the problems you do than to worry about the ones you leave undone.

- **You are not expected to tackle every single section!** Nor do you have to complete every single question in the sections you do tackle. The material is meant to be fun – as well as being challenging. So if you have to choose, concentrate on those items that interest you most. However, a problem will often not make sense unless you have understood earlier problems in the same section; so **make sure you complete the first few questions in each section** that you tackle.

- Many of the sections can be completed on your own. Some require you to talk to someone else about the ideas involved. It can often help to talk things through with someone else, even when it is not strictly needed in order to complete the task.

- Some sections may require you to visit a library or to use reference books. In addition there is a **Glossary** (that is, a mini-dictionary) at the end of the book which should help to explain any unfamiliar mathematical words or symbols which are used in the text.

- The **Comments and Solutions** are included to help you. Use them wisely! Don't peep too soon; but don't be afraid to use them to help you get unstuck. If you struggle to understand the Comments and Solutions for one question in a section, it will often help you to tackle other questions in that section.

  Remember that the most important thing is not just to get 'the answer. You should always present the method you use in a way that shows that it is correct. In mathematics this often means that a full solution to a problem has to use algebra!

1   *Solving and proving*

This activity focuses on:

- using algebra to solve simple word problems and to prove general results;
- thinking about the relationship between these two aspects of algebra.

Organisation/Before you start:

- Questions **1–3** assume you can solve linear equations; questions **4–6** assume you can multiply out brackets.

The problems should be tackled **in order** – starting with question **1**.

Solving

1   I am thinking of two numbers. My numbers are in the ratio 2 : 5. One number is 21 more than the other. What are my two numbers?

2   I am thinking of two numbers. My numbers are in the ratio 2 : 7. One number is 21 more than the other. What are my two numbers?

3   I am thinking of two other numbers which are again in the ratio 2 : 7. The product of the two numbers is 686. Can you tell me what my two numbers are?

If you managed all three questions easily, then you only need to read the next four paragraphs quickly. If you managed question **1** but struggled with questions **2** and **3**, then you should slow down.

Questions **2** and **3** may seem harder than question **1**, but from a mathematical viewpoint, the three problems are really very similar. That is because real mathematics is about general **methods**, and the general method behind all three problems is the same.

If you ignore general methods and just look for an answer, then question **1** may seem relatively easy. The numbers 2, 5 and 21 have been chosen in a particularly nice way (since 5 – 2 = 3 and 3 is a factor of 21), and this makes it possible to spot an answer. Even if you cannot spot an answer and decide just to guess, then it does not take long to come up with two numbers that work.

Unfortunately, when you guess an answer (or use trial and improvement), there is no way of knowing whether your guess is the **only possible** answer. So you cannot assume that the two numbers you come up with are in fact the required numbers. For example, if you tried to guess in questions **1–3** and managed to find two numbers, it is highly unlikely that your numbers were the ones I had in mind! Guessing is always allowed – and can often help you understand what needs to be done to solve a problem properly. But guessing is not mathematics. Moreover, the kind of answers you are likely to guess tend to be small integers or very familiar fractions. As soon as the answer to a problem is less familiar, then guessing is hopeless.

The only general mathematical way of solving problems involving unknown numbers is:

- to start by giving names to the unknown numbers

- then to write the given information in the form of equations.

So your solution to question **1** should begin:
Let the two numbers be x and y.

Proving

Make sure you read and understand the **Comments and Solutions** to questions **1–3** before tackling the remaining questions.

Proving general results using algebra is very like setting up and solving equations to find unknown numbers. We start with an example.

- The left-hand column in the table below contains the instructions.

- The central column contains a numerical example.

- The right-hand column follows the same instructions through, but in the **general** case – using symbols.

- The unexpected conclusion is stated underneath in the form of a **Claim** statement: the proof of this statement follows from the algebra in the right-hand column.

Instructions	Numerical example	Algebraic generalisation
Write down any three consecutive integers.	6, 7, 8	$a, a + 1, a + 2$
Work out the product of the first two integers.	$6 \times 7 = 42$	$a(a + 1)$
Work out the product of the last two integers.	$7 \times 8 = 56$	$(a + 1)(a + 2)$
Subtract the first product from the second.	$56 - 42 = 14$	$(a + 1)(a + 2) - a(a + 1)$ $= (a + 1)[(a + 2) - a]$ $= (a + 1) \times 2$

Claim: No matter which three consecutive integers you start with, the final answer is always equal to twice the middle integer.

Proof: If you follow the instructions using algebra, as in the right-hand column, you always get $2(a + 1)$, and $a + 1$ was the middle integer. **QED**

In each of the following questions, write out the corresponding **Claim** statement that is to be proved, and then use the same approach as in the above example to give the algebraic proof.

4 Write down any two consecutive **odd** integers. Show that their product is always 1 less than a multiple of 4.

5 Write down any three consecutive **odd** integers. Work out the square of the middle integer, and then subtract the product of the first and last integers. Prove that the difference is always equal to 4.

6 Write down any three consecutive integers.

    a Prove that the difference between the sum of the last two integers and the sum of the first two integers is always equal to 2.

    b Prove that the product of the first and the last integers is always one less than the square of the middle integer.

    c Prove that the difference between the product of the last two integers and the product of the first two integers is always equal to two thirds of the sum of all three integers.

7 Choose a two-digit integer. Reverse the digits to form another two-digit integer. Subtract the smaller two-digit integer from the larger. Prove that the answer is always a multiple of 9.

8 Choose a three-digit integer. Reverse the digits to form another three-digit integer. Subtract the smaller from the larger. Prove that the answer is always a multiple of 99.

The last two problems in this section are considerably harder – so don't feel bad about skipping them.

9 Find all triples of integers x, y, z whose sum is equal to their product.

10 Prove that the product of four consecutive positive integers can never be equal to a perfect square.

2 Congruent and similar

This activity focuses on:

- the classical criteria for congruence and similarity of triangles.

Organisation/Before you start:

- You need to know how to tell when two triangles are similar.

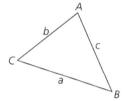

When you are given a triangle ABC, you know **six** separate pieces of information:

- the lengths of the three sides a, b, c, and

- the sizes of the three angles $\angle CAB$, $\angle ABC$, $\angle BCA$.

1 a Draw a triangle in which all three sides have different lengths. Label the vertices A, B, C. Measure the lengths a, b, c of the three sides BC, CA, AB respectively, and the sizes of the three angles $\angle CAB$, $\angle ABC$, $\angle BCA$. Mark your measurements on your diagram.

b Imagine that you need to give sufficient information to a friend by telephone so that she can draw an exact copy of your triangle. How many of the six pieces of information: a, b, c, $\angle CAB$, $\angle ABC$, $\angle BCA$ do you need to tell her?

If one triangle is 'an exact copy' of another, the two triangles are **congruent**. In question **1** you may have realised that you only need to give three of the six pieces of information in order to determine the triangle completely. However, you have to be careful which three pieces of information you give! Question **2** is designed to make you think which three pieces of information are enough, and which are not enough, to guarantee that two triangles are congruent.

Before starting question **2** it is convenient to introduce a useful shorthand.

- If the lengths of all three sides of a triangle are given, write this as *SSS* for short.

- If the sizes of all three angles in a triangle are given, write this as *AAA* for short.

- If the lengths of two sides and one angle are given, then there are two different possibilities: if the angle that is given is the angle **between** the two sides that are given, write this as *SAS* for short;

if the angle that is given is **not** between the two sides that are given, write this as *SSA*.

- If the sizes of two angles and the length of one side are given, then there are again two different possibilities: if the side that is given is the side **between** the two angles that are given, write *ASA* for short; if the side that is given is **not** between the two angles that are given, write *AAS* for short.

2   Suppose you give your friend three pieces of information about your triangle. Which of the six possible choices you can make

SSS, AAA, SAS, SSA, ASA, AAS

would leave her **most** confused (because there are infinitely many different triangles which seem to work)?

3   **a** Four of the choices in question **2** are sufficient to ensure that your friend's triangle will be congruent to yours. Which are they?

   b In order to determine a triangle, *SSA* is not enough. Find two triangles *ABC* and *A′B′C′* which are not congruent, even though

   $AB = A'B'$, $BC = B'C'$, and $\angle BCA = \angle B'C'A'$.

EXTRA 1

Prime numbers

Notice that the prime number $5 = 6 - 1$; the prime number $7 = 6 + 1$; and the prime number $11 = 2 \times 6 - 1$. Why is every prime number greater than 3, either 1 more or 1 less than some multiple of 6?

Solution: page 107

3 Surprising statistics

CHALLENGES

This activity focuses on:

• thinking more carefully about data.

Organisation/Before you start:

• You need to know about relative frequency and to be ready to think about frequency distributions.
• You need a local telephone directory.

1 Suppose everyone in your school wrote down the first (that is, left-hand) digit of their house number.

  a Which digit would not occur at all?
  How often would you expect the other nine digits to occur?

  b Now do the following experiment. Choose a page at random from your local telephone directory. Make a tally chart of the first digits of the house numbers that occur in the addresses listed on that page. Then work out the relative frequency with which each digit occurs as a first digit.

2 Before you carried out the experiment in question **1b** you might have thought that each digit would occur more or less as often as any other digit. Now you know differently! Try to explain what you discovered before going further.

In 1938 Frank Benford noticed what appeared to be a general law about certain kinds of **numerical data**. The phenomenon had also been noted by Simon Newcomb in 1881, but is now generally known as *Benford's Law*. Benford looked at sets of data from very different sources: areas of lakes, lengths of rivers, sizes of populations, and so on. For each dataset he made a tally chart of first **digits**. In each case he found a very similar **distribution** of the possible first digits: the **relative frequency** decreased as the digit increased. The digit 1 occurred with relative frequency roughly 0.3, while the digit 9 occurred with relative frequency roughly 0.05. Some typical results are shown in the **relative frequency chart** on the right.

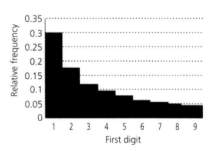

3 Use the glossary to make sure you understand the terms in **bold** type in the previous paragraph.

4 **a** Make a relative frequency chart for the data you found in question **1b**.

b How similar is your chart to the one shown above?

5 **a** Suppose there are exactly ninety-nine houses in a street, numbered 1–99. What would be the relative frequency of having first digit 1? What would be the relative frequency of having first digit 9?

b What if there were 999 houses in the street?

c How would these results change if there were 110, or 222 houses?

6 This unexpected phenomenon is what is known as *Benford's Law*. Suppose you are given a long list of random numbers whose first digits are distributed roughly according to Benford's Law. If you multiply all the numbers in your list by 2, you get a new list of numbers. Will your new list of numbers also satisfy Benford's Law (more or less)?

7 Around 1970 it was suggested that Benford's Law might be used to help detect firms who were returning false accounts. Does this sound realistic? If so, how might it be done?

EXTRA 2

Atoms in the universe

The 60 children leaving Hogwarts Junior School are lining up for their leavers' photo. They refuse to stand still and keep changing places. Which number do you think is biggest:

i the number of different ways 60 children can be arranged in one long line, or

ii the number of atoms in the universe?

Solution: page 108

4 *Right-triangle numbers*

This activity focuses on:

- reading and using information from a piece of written mathematics;
- generating Pythagorean triples;
- using algebra to formulate and prove general statements.

Organisation/Before you start:

- You need to be able to multiply out brackets and to collect terms.

Read the following passage (from Edme Mariotte, *Essay on logic* (Paris 1678, p. 645)). Then use it to answer questions **1–5**.

When the square of an integer is equal to the sum of the squares of two other integers, these three integers are called right-triangle numbers; one may propose the problem of finding a certain number – say four or five – of these right-triangles; having found by chance or otherwise, one of these triangles, such as 3, 4, 5; for 25, the square of 5 is equal to 16 and 9 together, which makes the squares of 4 and 3; one may notice that the greatest number 5 is composed of two squares; that is 4 and 1 of which 2 and 1 are the roots; that 3 is the difference of the same squares, & that the third number 4 is twice the product of the two roots 1 & 2. Following this observation one can take two other integers such as 3 and 2; and after having considered that 13 is the sum of the squares of these two integers, and that 5 is the difference of these same squares, one will see that if one takes from 169 (the square of 13), 25 (the square of 5), the result is 144, which is also a square number, the root of which is 12; and consequently that 13, 12 & 5 make a right-triangle in integers, and that 12 is twice the product of the roots 2 & 3. Similar observations can be made for two other integers such as 2 & 5; one will find that 29, made from their squares, 21 the difference of the same squares, & 20, twice their product, also form a right-triangle; for the square of 29, which is 841, equals the sum of 400 and 441, the squares of 20 and 21; whence one can conjecture that this rule is general, & that by this means one can find as many right-triangles as one may wish. Later we will seek the principles for demonstrating this rule.

1 What Mariotte calls *right-triangle numbers* are now known as **Pythagorean triples:** that is triples of positive integers – like 3, 4, 5 – such that the square of the largest is equal to the sum of the squares of the two smaller integers. If these integers are the lengths of the sides of a triangle, the triangle will be a right-angled triangle.

a How many Pythagorean triples does Mariotte mention in the above passage?

b Mariotte describes (rather badly!) a method for generating Pythagorean triples. Use his method to find a triple which is not mentioned in the above extract.

2 Mariotte describes how to generate a Pythagorean triple starting from any two integers – such as 1 and 2, or 3 and 2, or 5 and 2.

    a Write out as simply as you can a set of instructions for obtaining a Pythagorean triple starting from any two integers.

    b Does your set of instructions work for the starting pair 4 and 2? (Check to see whether the numbers it produces are integers and whether the squares of these three integers add up as they should.)

3 **a** Copy and complete the table on the right, which uses algebra to show that the method described by Mariotte really works.

    The method goes back to the Greek mathematician Diophantos (c. 200 AD).

    b Use Diophantos' method to find as many Pythagorean triples as you can with hypotenuse of length < 50.

Write down any two positive integers (roots).	a, b ($a > b$)
Square each integer.	a^2, b^2
Find the sum of the two squares.	$z = a^2 + b^2$
Find the difference of the two squares.	$y = a^2 - b^2$
Double the product of the two roots.	$x = 2ab$
The three integers x, y, z form a Pythagorean triple. (Check $x^2 + y^2 = z^2$.)	?

4 Find a Pythagorean triple with hypotenuse < 20 which is not produced by the method described by Mariotte.

5 **a** Write down any odd number $x > 1$. Square it. Find two consecutive integers y and z the difference of whose squares is equal to the square of your odd number. Show that x, y, z form a Pythagorean triple.

    b Prove that the procedure in part **a** produces a Pythagorean triple for every odd integer x.

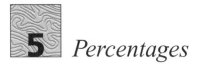

5 *Percentages*

This activity focuses on:

- understanding and extending work with percentages;
- using percentages in a variety of settings.

Organisation/Before you start:

- Questions **1–10** are meant to be done **without** a calculator.

1 a In a class, 10% of pupils score As, 20% score Bs, 30% score Cs, 40% score Ds. How many pupils score Es or worse?

   b In another class 10% of pupils score As, 15% score Bs, 20% score Cs, 25% score Ds and six score Es or worse. How many pupils are there in the class?

2 Which would you rather have: 10% of £30, or 30% of £10. Why?

3 Squeegees usually cost £20 each. My local hardware store is offering a discount of 10% off each one for a limited period, so I decide to buy three squeegees. How much will I pay altogether for the three squeegees?

4 When negotiating a big increase in pocket money you are given a choice: either an increase of 30% this year followed by a 40% increase next year, or an increase of 40% this year followed by a 30% increase next year. Which should you choose – and why?

5 What is 30% of 40% of 50% of £100?

6 A shopkeeper buys a coat for £39 and marks the price up by $33\frac{1}{3}\%$. The coat then fails to sell so is included in the next sale – where all prices are reduced by 25%. What price would you have to pay for the coat in the sale?

7 Alice finds one day that the price of a bowl of mock-turtle soup has suddenly increased by 50%. Next day the price increases by a further $33\frac{1}{3}\%$.

   a What total percentage increase has there been over the two days?

   b The price then drops by 50% the next day and $33\frac{1}{3}\%$ the day after that. What is the overall (net) percentage change over the four days?

c What if the first two price increases were as in part **a** but the price decreases in part **b** had been $33\frac{1}{3}$% on day 3 and 50% on day 4. What overall percentage change would there have been over the four days?

8 A meal costs £12.50. To that you have to add 17.5% VAT and a 12.5% service charge. Would it be cheaper to add the VAT first and then to add the service charge based on the resulting bill, or to add the service charge first and then to pay VAT on the total? Explain your answer.

9 **a** A shepherd loses 60% of her sheep and then finds 60% of those she lost. What percentage of her flock is still missing?

b Another shepherd loses three-quarters of his flock. Next day he finds all but 10% of the missing sheep. What percentage of his flock is still missing?

10 A school is laying two new football pitches – one for seniors and one for juniors. The senior pitch is to be 10% wider and 10% longer than the junior pitch.

a How much extra grass seed will be needed to sow the senior pitch (as a percentage of the amount needed for the junior pitch)?

b How much less grass seed will be needed to sow the junior pitch (as a percentage of the amount needed for the senior pitch)?

c How much extra white paint will be needed to mark all the white lines on the senior pitch (as a percentage of the amount needed for the junior pitch)?

11 The radius of a circle is increased by 20%.

a By what percentage does the perimeter increase?

b By what percentage does the area increase?

12 **a** By what percentage does the area of a circle increase if I double its radius?

b By what percentage does the radius of a circle increase if I double the area?

13 A cuboid has length 124 cm, width 69 cm and height 47 cm. Each of these lengths is increased by 50%.

a By what percentage does the volume of the cuboid increase?

b By what percentage does the surface area increase?

14 A cuboid has length 124 cm, width 69 cm and height 47 cm. Each of these lengths is to be increased by the same percentage $P\%$. What value of P should you choose if you want the volume of the cuboid to double?

15 Start with two identical circles $\mathcal{C}$ and $\mathcal{C}'$.

Change circle $\mathcal{C}$ by first doubling its radius and then halving its area to get a new circle $\mathcal{C}_1$.

Then change circle $\mathcal{C}'$ by first doubling its area and then halving its radius to get a new circle $\mathcal{C}'_1$.

Which of the three circles $\mathcal{C}$, $\mathcal{C}_1$, $\mathcal{C}'_1$ is biggest? Which is smallest?

16 Suppose the annual rate of inflation stays at 5% for 15 years. If a house costs £50 000 today, what would you expect the same house to cost in 15 years time?

EXTRA 3

Hoses

The fire brigade is about to start pumping flood water. They could use either one six-inch diameter pipe, or two four-inch diameter pipes. Which choice would allow them to pump away the water more quickly?

Solution: page 108

6 Divisibility rules

> ## This activity focuses on:
> - divisibility rules for prime factors;
> - using algebra to justify these divisibility rules.
>
> ## Organisation/Before you start:
> - You need to use simple algebra.

1 Write down any two-digit number '*ab*'.

Multiply the units digit *b* by 5. Add the answer to the tens digit *a*.

Rule: If the final answer is a multiple of 7, then so was the original number '*ab*'. If the final answer is not a multiple of 7, neither was the original number '*ab*'.

  a Try several different starting numbers – some which are multiples of 7 and some which aren't. Does the rule seem to work?

  b Consider a general two-digit starting number: '*ab*' = $10a + b$. The rule claims that:

  - if 7 divides $5b + a$, then 7 divides '*ab*' = $10a + b$; and
  - if 7 does not divide $5b + a$, then 7 does not divide '*ab*' = $10a + b$.

  Prove that this is correct.

  c Can you extend the rule to three-digit integers?

> '*ab*' = 83
> Units digit *b* = 3:
> ∴ 3 × 5 = 15
> Tens digit *a* = 8:
> ∴ 15 + 8 = 23
> 23 is not divisible by 7
> ∴ 83 is not divisible by 7

2 Write down another two-digit number '*ab*'.

Multiply the units digit *b* by 4. Add the answer to the tens digit *a*.

Rule: If the final answer is a multiple of 13, then so was the original number '*ab*'. If the final answer is not a multiple of 13, neither was the original number '*ab*'.

  a Prove that this rule is correct.

  b Extend the rule to three-digit integers.

> '*ab*' = 83
> Units digit *b* = 3:
> ∴ 3 × 4 = 12
> Tens digit *a* = 8:
> ∴ 12 + 8 = 20
> 20 is not divisible by 13
> ∴ 83 is not divisible by 13

3 **a** Find a similar rule for divisibility by 19. Prove that your rule is correct.

  b Does your rule extend to three-digit integers?

4 Find a similar rule for divisibility by 23. Prove that your rule is correct. Does your rule extend to three-digit integers?

7 *Fraction puzzles*

This activity focuses on:

- thinking more deeply about fractions;
- developing ideas of proof.

Organisation/Before you start:

- You need to be confident with the basic arithmetic of fractions.
- You need to work with algebra and inequalities.
- Do not use a calculator.

Products

1 a Work out the answers to these multiplications:

$$1\tfrac{1}{2} \times \tfrac{2}{3} = \underline{\hspace{1cm}} \; ; \; 2\tfrac{2}{3} \times \tfrac{3}{4} = \underline{\hspace{1cm}} \; ; \; 3\tfrac{3}{4} \times \tfrac{4}{5} = \underline{\hspace{1cm}} \; .$$

b Write down the product in this sequence which gives the answer 10.

c Write down the product in this sequence which gives the answer 49.

d State the general result for the nth term of the sequence and prove that it is correct.

Reciprocals

2 a Choose any positive integer.

Write down its reciprocal and subtract it from 1.

Write down the reciprocal of the result and subtract it from 1.

Keep repeating this process. What happens?

b Prove that what you noticed in part **a** always happens.

Fractionally larger

3 Start with a proper fraction $\tfrac{a}{b}$, where a and b are positive integers.

a Try particular values of a and b to decide which seems to be the larger of the two fractions: $\tfrac{a}{b}$ or $\tfrac{a+1}{b+1}$. Prove your claim.

b Which is the larger fraction: $\tfrac{a}{b}$ or $\tfrac{a-1}{b-1}$? Prove your claim.

c Which is the larger fraction: $\tfrac{a}{b}$ or $\tfrac{a+2}{b+2}$? Prove your claim.

d Which is the larger fraction: $\tfrac{a}{b}$ or $\tfrac{a+n}{b+n}$, where n is any positive integer? Prove your claim.

Smaller and greater

4 Let a, b, c, d be any positive integers such that $\frac{a}{b} < \frac{c}{d}$.

    a How big is the fraction $\frac{a+c}{b+d}$? Is it less than $\frac{a}{b}$? Or greater than $\frac{c}{d}$? Or does it lie between $\frac{a}{b}$ and $\frac{c}{d}$?

    b Prove your claim.

EXTRA 4

Tubes

An A4 sheet can be bent to make a cylinder in two different ways – either widthways or lengthways.

Which cylinder has the greater volume?

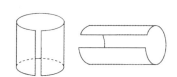

Solution: page 109

8 *Behold Pythagoras!*

This activity focuses on:

• Pythagoras' theorem.

Organisation/Before you start:

• You need to know what Pythagoras' theorem says.

Around 1150 AD the Hindu mathematician Bhaskara presented the two diagrams in **3** below as a proof of Pythagoras' theorem, with the single word 'Behold!'.

Look at each diagram in turn. Decide how it relates to Pythagoras' theorem. Then explain your conclusions to someone else, **with proofs**.

1

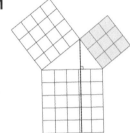

2

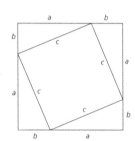

3

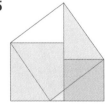

4

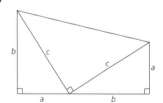

5

6

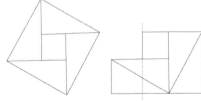

7

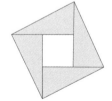

9 *Not what you expect*

This activity focuses on:

• working systematically and accurately;
• identifying and testing possible patterns in the first few terms of a sequence;
• making general conjectures where possible and proving them.

Organisation/Before you start:

• If you are working with a partner, work on the same question – then discuss and agree on the solution.

1 You have a large supply of white beads and black beads which can be threaded onto a string to make different necklaces.

• A necklace with one bead can be made in two ways – either one black bead, or one white bead.

• A necklace with two beads can be made in three ways – both black, both white, or one black and one white.

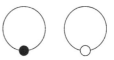

a There are four different necklaces with three beads. Find them all.

b Draw up a table which lists the number of different necklaces that can be made with four and five beads.

c Does this information give you any idea what to expect for six beads? Now work systematically to find all possible different necklaces with six beads.

2 There are five ways of making a total of 4 using positive whole numbers:

4; 3 + 1; 2 + 2; 2 + 1 + 1; 1 + 1 + 1 + 1.

a Draw up a table of the number of different ways of making each total from 1 up to 5.

Total	Number of ways
1	?
2	?
3	?
4	5
5	?
6	?
7	?

b Does the information in part **a** give you any idea about what to expect for 6 and 7?

Now work systematically to find the number of different ways of making 6 and 7.

3 In climbing stairs you can take either one step or two steps at a time. Suppose that you can combine these short (one step) and long (two step) steps in any way you like; then there are five different ways to climb a four-step staircase:

$$1 + 1 + 1 + 1;\ 1 + 1 + 2;\ 1 + 2 + 1;\ 2 + 1 + 1;\ 2 + 2.$$

a Draw up a table showing the number of ways of climbing staircases with from 1 to 5 steps.

b Does the information in part **a** suggest what to expect for staircases with 6 or 7 steps? Make a prediction, then work systematically to see whether you were right.

c State a general conjecture about how you think the sequence continues. Then determine – with proof – whether your guess is correct.

Total	Number of ways
1	?
2	?
3	?
4	5
5	?
6	?
7	?

4 How many different ways are there of writing the positive integer 10 as a sum of 1s and 2s only if the order of the terms matters (so $1 + 1 + 2 + 2 + 2 + 2$ is different from $1 + 2 + 1 + 2 + 2 + 2$)?

5 You have a large supply of 1p and 2p stamps and want to know how many different ways there are to make up Np.

a • You can make 1p in just one way.

  • You can make up 2p in exactly two different ways (2 and 1 + 1).

  • You can make up 3p in exactly two different ways (2 + 1 and 1 + 1 + 1).

Copy and complete the table to find the number of different ways each total stamp value can be made up.

b State a general rule for the number of different ways Np can be made using 1p and 2p stamps. Then prove that your rule is correct.

Total	Number of ways
1p	1
2p	2
3p	2
4p	?
5p	?
6p	?
7p	?

6 a You have a large supply of 1p, 2p and 3p stamps.

Copy and extend the table to find the number of different ways each total stamp value up to 7p can be made up.

b Did you get the answers you expected for 6p and for 7p?

Try to find a general rule for the number of different ways of making Np using 1p, 2p and 3p stamps.

Total	Number of ways
1p	1
2p	2
3p	3
4p	?

7 a Suppose you want to know in how many different ways you can write a given positive integer n as a sum of one or more parts where the order matters (so $1 + 2$ and $2 + 1$ are different).

- When $n = 1$, you can only write $1 = 1$, so there is 1 way.

- When $n = 2$, you can write $2 = 2$ or $2 = 1 + 1$, so there are 2 ways.

How many different ways are there of writing the positive integer 3 as a sum of one or more parts?

How many different ways are there of writing the positive integer n as a sum of one or more parts?

b Suppose you mark n points on the circumference of a circle – **not** equally spaced.

Points	Number of regions
1	1
2	2
3	?
4	?
5	?
6	?

When you draw in all possible chords, the inside of the circle is divided into regions.

- If $n = 1$, you get 0 chords; $\therefore$ 1 region.

- If $n = 2$, you get 1 chord; $\therefore$ 2 regions.

- If $n = 3$, you get 3 chords; $\therefore$ ____ regions.

What is the largest possible number of regions inside the circle with 6 points?

c The Big-Enders and the Little-Enders are contesting n seats in the Lilliput election. Suppose you want to know the number of different ways the Big-Enders can win an **odd** number of seats.

- If the number of seats contested is $n = 1$, the possible results are:

    B (Big-Enders win the one seat), and

    L (Little-Enders win the one seat).

So there is just one **odd** result for the Big-Enders, namely B.

- When $n = 2$, the possible results are

    $BB, BL, LB, LL,$

so there are just 2 **odd** results for the Big-Enders, namely BL and LB.

If the Big-Enders and the Little-Enders contest n seats, in how many different ways can the Big-Enders win an **odd** number of seats?

10 *Onwards and upwards*

This activity focuses on:

- identifying trends in data;
- using the identified trends to make predicitions.

Organisation/Before you start:

- You need to know about scattergrams and lines of best fit.

In 1896 Baron Pierre de Coubertin created a modern version of the ancient Greek tradition of holding Olympic Games. The modern Olympic Games have been held every four years since 1896 (except that there was an extra games in 1906, and there were no games held during the war years of 1916, 1940 and 1944).

The data for the performance of the gold medal winners in the men's high jump, long jump and discus are given below. The height of the winning high jump, the length of the winning long jump, and the distance thrown by the winner in the discus are given in centimetres.

Year	High jump	Long jump	Discus
1896	181	635	2915
1900	190	718	3604
1904	180	734	3928
1908	191	748	4089
1912	193	760	4521
1920	194	715	4468
1924	198	744	4615
1928	194	773	4732
1932	197	763	4949
1936	203	806	5048
1948	198	782	5278
1952	204	757	5503
1956	212	783	5636
1960	216	812	5918
1964	218	807	6100

Year	High jump	Long jump	Discus
1968	224	890	6478
1972	223	824	6440
1976	225	835	6750
1980	236	854	6664
1984	235	854	6660
1988	237	872	6881
1992	234	870	6512

1 Make three scattergrams – one for each event, containing all the given data for that event.

2 Describe the overall shape of the distribution revealed by each scattergram.

3 Use the scattergram for each event to predict the likely winning performance at the Olympic Games in: **a** 1996; **b** 2000.

Then check to see how close your prediction was to the actual winning performance.

EXTRA 5

Rhino

A female two-horned rhino weighs 1480 kg and is 315 cm long. Her baby is 76 cm long. How much would you expect the baby rhino to weigh?

Solution: page 109

11 *Very* small

This activity focuses on:

• developing the use of standard form with negative indices.

Organisation/Before you start:

• You will need a scientific calculator.
• You will need access to scientific reference books.
• If you are working with a partner, pick some questions to work on individually. Then discuss and compare your methods and your answers.

All answers should be given in standard form.

1 Protozoa

A single-celled protozoa is about one tenth of a millimetre across. Write its diameter in metres.

2 Water molecules

A molecule of water weighs roughly 3×10^{-23} kg.

Each living human body is roughly 65% water.

Roughly how many molecules of water are there in your own body?

3 The thickness of paper

How thick is a single sheet of ordinary paper?

4 Sizes

Find out how big is:

a a typical grain of pollen

b an average virus

c a hydrogen atom.

5 The scale of things

Here is a length scale indicating **metres** in powers of 10. It shows familiar objects of approximately the correct size. Find examples of other familiar objects which continue the scale as far as you can in both directions.

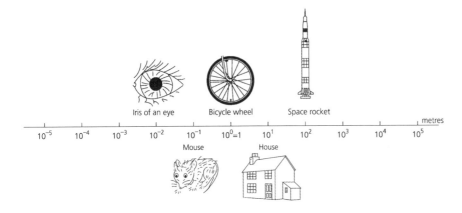

6   A teaspoon of oil

In 1770 Benjamin Franklin did an experiment on Clapham
Common. On a very still day, when the water on a pond was very
calm, he gently emptied one teaspoon of oil onto the surface of the
water. The oil spread out until it covered one acre of the surface.
Assuming that the layer of oil spread out until it was just one
molecule thick, what does this experiment suggest as a good
estimate for the diameter of one molecule of the oil used?
(1 acre ≈ 0.405 hectares; 1 teaspoon ≈ 5 ml.)

EXTRA 6

Three card trick

You have three cards:

- one is black on both sides

- one is white on both sides

- one is black on one side and white on the other.

a   If you put the cards down beside each other on the table,
    what is the probability that you see one black and two white
    sides?

b   You pick a card at random and see that it is black on one side.
    What is the probability that it is black on the other side also?

c   If you pick two of the three cards at random and lay them
    down on the table, what is the probability that you see one
    black and one white?

Solution: page 110

Very small

12 *Fractions and decimals*

This activity focuses on:

- understanding and working with recurring and terminating decimals.

Organisation/Before you start:

- Do not use a calculator.

Three starters

1. **a** Write down three fractions which are greater than 0.3 and less than 0.5. Give your answers as fractions in their simplest form.

    b How did you choose your fractions in part **a**? How do you know that your fractions lie between 0.3 and 0.5?

2. Write down three fractions which are greater than $\frac{1}{2}$ and less than $\frac{3}{4}$.

3. Is $\frac{7}{10}$ closer to $\frac{3}{4}$ or to $\frac{2}{3}$? How can you be sure?

From decimals to fractions and from fractions to decimals

There are two different ways to approach the first three questions. The first is to change everything to fractions and work solely with fractions; the second is to change everything to decimals and work solely with decimals. You should be happy with both methods and understand which is the most appropriate in any given setting.

A decimal which stops (or which **terminates**) can always be changed to a **decimal fraction** (that is, a fraction with denominator equal to a power of 10):

$$0.1 = \frac{1}{10}, \ 0.01 = \frac{1}{100}, \ 0.001 = \frac{1}{1000}, \text{ and so on.}$$

$$\therefore \ 0.234 = 0.2 + 0.03 + 0.004$$
$$= \frac{2}{10} + \frac{3}{100} + \frac{4}{1000}$$
$$= \frac{234}{1000}$$
$$= \frac{117}{500}.$$

4. Write down the fractions (in simplest form) for these terminating decimals.

    a 0.25   **b** 0.24   **c** 1.25   **d** 0.625   **e** 0.5625   **f** 1.234

There are two ways to turn a fraction – such as $\frac{1}{8}$ – into a decimal.

• The first is by doing the division:

$$8\overline{)1 . {}^1 0 {}^2 0 {}^4 0} 0 . 1 2 5$$

• The second is by writing the fraction as a decimal fraction (by multiplying both numerator and denominator to make the denominator a power of 10):

$$\frac{1}{8} = \frac{1}{2^3} = \frac{1 \times 5^3}{2^3 \times 5^3} = \frac{5^3}{10^3} = \frac{125}{1000}.$$

5 Work out the decimals for each of these fractions using **both** methods – by doing the division **and** by making the denominator a power of 10.

a $\frac{1}{4}$ **b** $\frac{7}{50}$ **c** $\frac{9}{8}$ **d** $\frac{7}{20}$ **e** $\frac{13}{25}$ **f** $\frac{5}{32}$

The second of these methods for turning a fraction to a decimal depends on **finding a multiplier** which will turn the denominator into a power of 10. This can be done for the fraction $\frac{1}{8}$; but for most fractions it is impossible. Here is the reason why.

The only prime factors of 10 are 2 and 5, so these are also the only prime factors of any power of 10.

However, if you start with a fraction like $\frac{1}{3}$ and multiply by $\frac{k}{k} = 1$, you get the same fraction $\frac{1}{3}$ written in the new form $\frac{k}{3k}$. So the denominator still has 3 as a prime factor. Therefore the denominator can never be equal to a power of 10.

The only way to find the decimal for a fraction like $\frac{1}{3}$ is to carry out the division.

6 Work out the **exact** decimal for each of these fractions by doing the division.

a $\frac{1}{3}$ **b** $\frac{1}{6}$ **c** $\frac{1}{7}$ **d** $\frac{1}{9}$ **e** $\frac{1}{11}$ **f** $\frac{3}{7}$

Where the denominator of a fraction cannot be made an exact power of 10, its exact decimal will never terminate – so it must go on forever.

Whenever the decimal of a fraction goes on forever, it is impossible to write it out in full. Fortunately, it always ends with a single block of digits that **recur**; the exact decimal is then written by marking the beginning and end of the block of digits that recurs with a dot – like this:

$$\frac{1}{3} = 0.333\,333\ldots\text{ (forever)} = 0.\dot{3}$$

$$\frac{1}{6} = 0.166\,666\ldots\text{ (forever)} = 0.1\dot{6}$$

$$\frac{1}{28} = 0.035\,714\,285\,714\,285\,714\,28\ldots\text{ (forever)} = 0.035\dot{7}14\,2\dot{8}.$$

Notice that the recurring block does not always begin immediately after the decimal point.

In question **4** you saw how easy it is to turn a **terminating** decimal into a fraction. In fact any **recurring** decimal corresponds to a fraction – but it is not at all clear which fraction it corresponds to: you may recognise $0.\dot{6}$ (as $\frac{2}{3}$), or even $0.1\dot{6}$ (as $\frac{1}{6}$); but you may not recognise $0.8\dot{3}$, and you may never have seen $0.1\dot{2}\dot{3}$ before in your life.

For fractions like these there is an easy – and very clever – trick which makes it possible to find the corresponding fraction:

• Call the unknown fraction x.

• If the recurring block consists of **just one digit**, multiply by 10 to get $10x$;

  then subtract x from $10x$ to get $10x - x = 9x$;

  finally divide the answer by 9 to find x as a fraction.

For example: Let $x = 0.1\dot{6}$.

$\therefore\ 10x\quad = 1.\dot{6}.$

$\therefore\ 10x - x = 1.666\,666\ldots\text{ (forever)} - 0.166\,666\,6\ldots\text{ (forever)}$

$\qquad\qquad = 1.6 - 0.1$

$\qquad\qquad = 1.5$

$\therefore\qquad\quad 9x = 1.5 = \frac{3}{2}$

$\therefore\qquad\quad\ x = \frac{3}{18} = \frac{1}{6}.$

If the recurring block consists of **two** digits (as for the recurring decimal $0.1\dot{2}\dot{3}$), you have to multiply x by 100, and then subtract x to get $99x$.

7 Use this method to find the fractions corresponding to these recurring decimals.

a $0.\dot{5}$ **b** $0.8\dot{3}$ **c** $0.\dot{1}\dot{5}$ **d** $0.0\dot{9}$ **e** $0.1\dot{2}\dot{3}$ **f** $0.\dot{1}2\dot{3}$

Why do fractions have recurring decimals?

Suppose you carry out the division to calculate the decimal for $\frac{2}{11}$:

$$
\begin{array}{r}
0 \ . \ 1 \ 8 \ 1 \\
11\overline{)2 \ . \ 0 \ 0 \ 0 \ 0 \ 0 \ 0 \ \ldots}
\end{array}
$$

It is tempting to believe that the decimal recurs because the digits in the answer **look as though** they are going to recur forever. But such optimism has no logical basis: you have no reason to know whether a decimal that begins

$$0.18 \ldots , \text{ or } 0.181 \ldots , \text{ or } 0.1818 \ldots$$

will continue repeating 18s forever. Indeed it is highly unlikely that such a wild guess – based only on the first few decimal places – will be correct.

The lesson here is that, when you do a division, the first few decimal digits in the answer tell you **nothing** about how the decimal continues.

So how can we be sure that the decimal for $\frac{2}{11}$ really is equal to $0.\dot{1}\dot{8}$? The important mathematics here lies **not** in the output of the division procedure (above the line), but in the ∗e∗ai∗∗e∗∗.

8 **a** What does ∗e∗ai∗∗e∗∗ stand for?

b Repeat the division to calculate the decimal for $\frac{2}{11}$, filling in the ∗e∗ai∗∗e∗∗ as you go (below the line).

$$
\begin{array}{r}
0 \ . \ 1 \ 8 \ 1 \\
11\overline{)2 \ . \ {}^{2}0 \ 0 \ 0 \ 0 \ 0 \ 0 \ \ldots}
\end{array}
$$

c What do you notice about the third ∗e∗ai∗∗e∗? Why does this guarantee that the digits **in the answer** must recur?

9 Carry out the division for each of these fractions, filling in the remainders as you go. How far do you have to go before you **know** that the answer recurs? Explain how you know.

a

$$
\frac{3}{7}; \quad
\begin{array}{r}
0 \ . \ 4 \ . \ . \ . \ . \ . \ . \\
7\overline{)3 \ . \ {}^{3}0 \ 0 \ 0 \ 0 \ 0 \ 0 \ 0 \ \ldots}
\end{array}
$$

b $\frac{4}{13}$;

$$
13 \overline{)4 \,.\, {}^{4}0\ 0\ 0\ 0\ 0\ 0\ 0}\ \ldots \quad = 0\,.\,3\,.\,.\,.\,.\,.\,.
$$

c $\frac{1}{19}$;

$$
19 \overline{)1 \,.\, {}^{1}0\ {}^{10}0\ {}^{5}0\ 0\ 0\ 0\ 0\ 0\ 0}\ \ldots \quad = 0\,.\,0\ 5\,.\,.\,.\,.\,.\,.
$$

10 Suppose you are challenged to work out the exact decimal for some new fraction (such as $\frac{1}{23}$, or $\frac{1}{32}$, or $\frac{1}{89}$). What do you know about the sequence of remainders that guarantees that the decimal will either terminate or will eventually recur?

EXTRA 7

Father Christmas

Father Christmas has two children.

a   If I tell you that at least one of the children is a boy, what is the probability that the youngest child is a boy?

b   If I tell you that the older of the two children is a boy, what is the probability that the youngest child is also a boy?

Solution: page 111

13 *Pigeons and pigeon-holes*

This activity focuses on:

• understanding and using an unfamiliar counting principle.

The **pigeon-hole principle** states:

Suppose a number of pigeons fly into pigeon-holes.
If there are more pigeons than pigeon-holes,
then some pigeon-hole must receive more than one pigeon.

The example below shows how the pigeon-hole principle can be used to solve a simple problem.

Problem: Sadiq has two kinds of socks: red socks and green socks. The socks in his sock drawer are all mixed up; none of them are in pairs. If Sadiq chooses with his eyes shut, how many socks must he choose to be sure of having at least one pair of the same colour?

Solution:
• The first two socks chosen could be of different colours.
So Sadiq has to choose at least three socks if he wants to be sure of choosing a pair.

• Imagine two pigeon-holes labelled 'red' and 'green', and think of the socks which Sadiq chooses as the 'pigeons' to go in the pigeon-holes. You know that there are just two pigeon-holes. So as soon as there are at least three socks, some pigeon-hole will receive at least two socks – which will be a pair of socks of the same colour.
Hence Sadiq only needs to choose three socks to be sure of having a pair. **QED**

1 a Sandra is getting dressed in a hurry – and in the dark. Her sock draw contains socks of three different colours, none of them in pairs. How many socks must she take from her sock drawer to be sure of having at least one pair of the same colour? Use the pigeon-hole principle to justify your answer.

b What if Sandra had socks of four different colours? How many must she choose to be sure of having at least one pair of the same colour? What if Sandra had socks of n different colours?

2 a If two integers sum to 100, one of them must be ≥ 50. Use the pigeon-hole principle to prove this obvious fact.

b If five integers sum to 100, one of them must be ≥ 20. Use the pigeon-hole principle to prove this fact.

3 Each human head has at most a million hairs on it. Use this to prove that there are at least two heads in London with the same number of hairs.

4 **a** In Hogwarts School, every pupil knows at least one other pupil. Prove that there must be at least two pupils who have the same number of friends.

b Suppose you are not told that every pupil knows at least one other pupil. Is it still true that there must be at least two pupils who have the same number of friends?

5 A football league has n teams.

a If each team plays every other team once, can you be sure that at the end of the season there will be two teams who have drawn exactly the same number of games?

b What if during the season each team plays every other team twice – once at home and once away from home? Can you still be sure that at the end of the season, there will be two teams who have drawn exactly the same number of games?

6 **a** Here is a claim found in a puzzle book.

> • Pick any ten integers < 100.
> • Then you will always be able to find two different subsets of these ten integers which have the same sum.

Try it for yourself. Does it seem to be true?

b Choose three integers a, b, c < 100. If you want to choose a subset of these three integers, there are three ways to choose just one number, three ways to choose just two numbers, and one way to choose all three of the numbers. So there are seven different subsets you could choose. Prove that either some subset has sum which is a multiple of 7, or you can find two subsets whose sums differ by a multiple of 7.

c If you choose four integers a, b, c, d, how many different subsets of these four integers are there?

d If you choose ten integers, how many different subsets of these ten integers are there?

If each of your ten integers is < 100, explain why none of these subsets can have sum ≥ 1000. Use this to prove the claim in the box in part **a**.

14 *A short history of* π

This activity focuses on:

- comprehension of passages related to π;
- appreciating part of the history of π.

Organisation/Before you start:

- If you are working with a partner, pick some questions to work on individually. Then discuss and compare your conclusions.

Circles have been recognised as being important for millennia – and not just because of the wheel. Several ancient civilisations realised that when you double the diameter of a circle the length of the circumference also doubles. However, the ancient Greeks were the first to **prove** (around 300 BC) that the ratio

$$\text{\textit{circumference of circle : diameter of circle}} \qquad (1)$$

is constant. Before this the concept of π as a universal constant did not really exist.

For two thousand years after it was proved that the ratio (1) was constant, this constant did not have a simple name. The Englishman William Jones, a friend of Isaac Newton, was the first to use the symbol π to denote this constant ratio in a textbook he published in 1706. But the use of the symbol did not really catch on until it was adopted in the late 1730s by Leonhard Euler – the greatest mathematician of the eighteenth century.

1   Shortly before William Jones first used the symbol π with its modern meaning, William Oughtred and others denoted the ratio

$$\text{\textit{circumference : diameter}}$$

by $\frac{\pi}{\delta}$. Why do you think Oughtred used π and δ?

2   Old texts have to be read from the perspective of their own time. So even where you find indications of a practical rule-of-thumb which relates the circumference of a circle to its diameter, you should not assume that those who wrote the text were thinking about the modern constant π. For example, the Old Testament contains this description of part of the temple that Solomon built (1 Kings, Chapter 7, verse 23; also 2 Chronicles, Chapter 4, verse 2):

> *Then he made the molten sea; it was round, ten cubits from brim to brim, and five cubits high, and a line of thirty cubits measured its circumference.*

a This passage – dating from before 550 BC – describes some very large circular container (containing 'the molten sea'). What is given as the diameter of the container? What is given as its circumference?

b What does this suggest as the accepted ratio

circumference : diameter ?

3 One of the oldest estimates of the number we now call π is given in the Rhind papyrus. This was written in Egypt by Ahmes the scribe around 1700BC. The papyrus contains many problems that were probably used to teach mathematics. In Problem 50 a circular field with diameter 9 khet is given and its area is to be calculated. A khet was equal to 100 royal cubits, or approximately 50 metres. Here is Ahmes' solution.

> *Subtract $\frac{1}{9}$ of the diameter, namely 1 khet.*
> *The remainder is 8 khet.*
> *Multiply 8 by 8.*
> *It makes 64.*
> *Therefore it contains 64 setat of land.*

a What is a **setat**?

b What was the diameter in metres of the circular field in Problem 50? Do you think the problem is about a real field? Or were the numbers chosen to make the calculation easy?

c Ahmes works with the diameter $D = 2r$ instead of the radius r. What does the formula πr^2 become if you write it in terms of the diameter D? Now use this formula and the value of π in your calculator to work out the correct area of the circular field in Problem 50.

d Ahmes did not know about π, and did not know the formula πr^2. Instead he used an approximate formula.
He first replaced the diameter D by $\frac{8}{9}D$, and then squared the result.
So he used the approximate formula $(\frac{8}{9}D)^2$ for the area.
What value of π is this equivalent to?

4 At about the same time as Ahmes was writing his papyrus in Egypt, the Babylonians calculated the circumference of a circle using the inscribed regular hexagon. It seems they knew that each side of the inscribed regular hexagon has length equal to the radius of the circle. (This was one reason why they divided a complete turn into 360 degrees – the same angle measure we use today.)

a In 1936 a tablet was excavated about 200 miles from Babylon. It had on it several geometrical drawings and was dated at around 1600 BC. This tablet states that the ratio

perimeter of hexagon : circumference of circle $= \frac{57}{60} + \frac{36}{60^2}$.

Use this information to show that the Babylonian method is equivalent to taking $\frac{3}{\pi} = \frac{57}{60} + \frac{36}{60^2}$.

b Simplify the equation in part **a** to find a Babylonian value of π.

5 As indicated at the beginning of this section, the ancient Greeks were the first to **prove** that the ratio

circumference : diameter

has the same value for every circle. So they were the first people who could give this ratio a fixed name and begin to pin down its exact value.

Around 230 BC Archimedes observed that the regular hexagon inscribed in a circle of diameter $D = 2r$ has six sides, each of length r, so has perimeter $3D$.

Since the hexagon is inside the circle, the perimeter of the hexagon is less than the circumference of the circle:

∴ $3D < \pi D$.
∴ $3 < \pi$.

a Archimedes then bisected each side of the regular hexagon to obtain a regular 12-gon inscribed in the circle.
Find the length s of each side of the inscribed regular 12-gon.
Then use the fact that $12s < \pi D$ to get a better estimate for π.

b Archimedes then considered the regular hexagon outside the circle whose sides just touch the circle.
Find the length t of each side of this regular hexagon.
Then use the fact $6t > \pi D$, to show that $\pi < 2\sqrt{3}$.

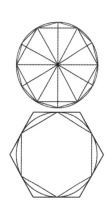

Archimedes then used 24-gons, 48-gons and finally 96-gons inside and outside the circle to prove that

$$3\tfrac{10}{71} < \pi < 3\tfrac{1}{7}.$$

Other cultures used different approximations for π – though it is not always clear how they were obtained.

6 Liu Hui in China around 260 AD used a value close to 3.1416.
Around 480 AD Zu Chongzhi wrote that $\frac{355}{113}$ was a better approximation than $\frac{22}{7}$. How accurate is the approximation $\frac{355}{113}$?

15 *Interesting! But why does it work?*

This activity focuses on:

• presenting some simple numerical tricks for you to do, and to understand.

Organisation/Before you start:

• This section should be done with a partner. Try out each trick; then find out why it works.

1 Ask your partner to hide a 1p coin in one hand and a 2p coin in the other hand. Tell them to triple the value of the coin in their right hand and add this to double the value of the coin in their left hand. All they tell you is the answer. *If the answer is even, the 2p coin is in their right hand. If the answer is odd, the 2p coin is in their left hand.*

    a How does this work?

    b Would the same rule work if you used a 2p coin and a 5p coin? Or a 2p coin and a 10p coin? Or a 5p coin and a 10p coin?

2 Decide who is to be **the magician**, and who is to be **the apprentice**.

    • The magician turns away while the apprentice rolls three ordinary dice on the table. The apprentice then adds the three scores and remembers the first total. The apprentice then picks up one of the dice, looks at the number on the bottom and adds this to the first total to get the second total. Finally, this one dice is rolled again and the number it shows is added to the second total to obtain the *final score.*

    • The magician is told none of the three totals. However, as soon as the magician turns round and sees the three dice on the table, she can immediately announce the final score. How does she do this?

3 You turn away while your partner rolls three dice. Without peeping, you ask your partner to:

    • multiply the number on the first dice by 2

    • add 5 to the answer

    • multiply the result by 5

    • add to this the number showing on the second dice

    • multiply the answer by 10

- finally add the number on the third dice, and tell you the final total.

You can then tell your partner which three numbers he rolled. How do you do this? And why does it work?

4 Ask your partner to choose a page in a book, a line on that page, and a word on that line. Your partner should record and keep secret:

- the page number

- the line number (counting from the top of the page)

- the word number (counting from the left-hand end of the line it is in).

You now ask your partner to:

- double the page number

- multiply the answer by 10

- multiply the result by 5

- add the line number

- add 5 to the total so far

- multiply the result by 100

- add the number of the word on the line

- add 611 to the result, and announce the final answer.

You should now be able to find the exact word in the book. How? And why does it work?

16 *Area puzzles*

CHALLENGES

This activity focuses on:

• thinking more deeply about simple geometrical ideas.

Organisation/Before you start:

• You need to know about the area of a circle and similarity of triangles.

1 A quadrilateral has sides of length 3, 4, 5 and 12. What is its area?

2 The diagram shows three identical squares.

In each square a shaded region is obtained by drawing certain circular arcs.

Which diagram has the largest shaded region?

3 The diagram shows two overlapping circles. The two non-overlapping regions have areas A and B. As the area of overlap changes, the values of A and B also change.

Prove that, no matter how big or small the overlap is, the difference $A - B$ always stays the same.

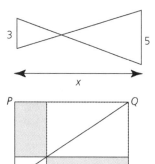

4 Calculate the area of this lop-sided bow-tie shape in terms of x.

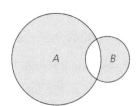

5 The diagram shows a rectangle *PQRS*.

Which of the two shaded areas is biggest?

6 Area is usually measured using 1 by 1 squares as the unit of area. This question uses '1 by 1 equilateral triangles': each 1 by 1 equilateral triangle is called a **triangular unit**, or t.u. for short. To find the area of a shape (in t.u.s), count the number of 1 by 1 equilateral triangles that fit inside it.

a Draw equilateral triangles with sides of length 1, 2, 3, and so on on an equilateral triangular grid.

Find a formula for the area (in t.u.s) of an equilateral triangle with sides of length a. Explain why your formula is correct.

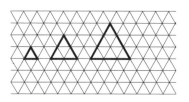

maths CHALLENGE 3   36

b Draw parallelograms with angles of 60° and 120° and with sides running along the grid lines, on an equilateral triangular grid.

Find a formula for the area (in t.u.s) of such a parallelogram with sides of lengths a and b.

c Draw triangles with one angle of 60°, and with two sides running along the grid lines, on an equilateral triangular grid.

If the two sides meeting at the 60° angle have lengths a and b, find a formula for the area (in t.u.s) of the triangle.

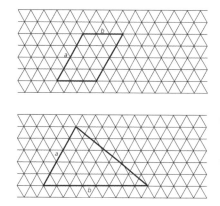

d Extend the formulas you found in parts **a**–**c** to give the area **in triangular units** of any triangle with a 60° angle and any parallelogram with a 60° angle.

17 A sequence of triples

This activity focuses on:

• spotting numerical patterns and generalising them algebraically.

Organisation/Before you start:

• You should tackle Challenge 16 in Book 2 before this section.

1 The table on the right shows the first three triples (a, b, c) in a sequence of triples.

	a	b	c
1st triple	3	4	5
2nd triple	5	12	13
3rd triple	7	24	25
4th triple			
5th triple			
6th triple			
....			
....			

   a Look at the entries in the first column – the 'a' column. What do you expect the next two entries in this column to be? Explain how each entry is obtained from the entry before it. Give a formula which expresses the nth entry in the 'a' column in terms of n.

   b Look at the entries in the second column – the 'b' column. What do you expect the next two entries in this column to be? Explain how each entry is obtained from the entry before it.

   c Look at the entries in the third column – the 'c' column. What do you expect the next two entries in this column to be? Explain how each entry is obtained from the entry before it.

2 **a** The first triple (3, 4, 5) should look familiar. What do you know about the sum of the squares of the first two numbers in this triple?

   Does the second triple (5, 12, 13) have this same property?

   Does the third triple (7, 24, 25) have the same property?

   b Does your guess for the fourth triple have this property? How about your guessed fifth triple?

3 In question **1a** it was fairly easy to write a formula in terms of n for the nth entry in the first column.

   a For the entries in the second column, in question **1b** you were only asked to explain how to work out each entry from the entry before it. Now find a formula (in terms of n) for the nth entry in the second column – that is, for the b in the nth triple (a, b, c).

   b For the entries in the third column, in question **1c** you were only asked to explain how to work out each entry from the entry before it. Now find a formula in terms of n for the nth entry in the third column – that is, an formula for c in the nth triple (a, b, c).

c Use your formulae for the integers a, b, c in the nth triple (a, b, c) to check that $a^2 + b^2 = c^2$.

4 The numbers in the sequence of triples (a, b, c) are related in lots of ways.

a Notice that in the first triple, $3^2 = 4 + 5$. Does the same relation $a^2 = b + c$ hold for the next few triples too? Does this relation hold for the nth triple?

b Notice that in the first triple (a, b, c),

$$a + b + c = 3 + 4 + 5 = 12 = 3 \times 4 = 4a.$$

Work out $a + b + c$ for the second triple and factorise the answer. Guess a formula for the sum $a + b + c$ of the three entries in the nth triple. Use algebra to prove that your guess is correct.

5 Suppose you wanted to find a new triple (a, b, c) **between** the first triple and the second triple. What would be the value of a? What would be the value of b? What would be the value of c?

EXTRA 8

Perfect logic

Alice, Becky and Claire are all perfectly logical. Each can make perfect deductions – and each of them knows that the others can too. The three of them are shown seven stamps: two red stamps, two green stamps, and three yellow stamps. They are then blindfolded and one stamp is stuck on the forehead of each girl: the remaining four stamps are hidden in a drawer. The blindfolds are removed and Alice is asked: 'Do you know anything – either positive or negative – about the colour of the stamp on your own forehead?' She replies 'No.' Becky is then asked the same question, and she also replies 'No.' Claire immediately breaks into a smile – because she now knows the colour of the stamp on her forehead. What colour must it be, and why?

Solution: page 111

A sequence of triples

18 *Folding squares, equilateral triangles and regular hexagons*

This activity focuses on:

• simple applications of congruence in the context of folding.

Organisation/Before you start:

• You will need a few sheets of plain A4 paper.

1 A sheet of A4 paper is rectangular. For many paper-folding constructions (for example, in origami) you have to start with a **square**.

 a Take an A4 sheet of paper. Find out how to use folding to make a perfect square (without measuring).

 b Explain why your method produces a perfect square.

• Once you have folded a perfect square on one end of your A4 sheet, cut off the spare rectangle from the other end.
You now have a perfect square *PQRS*.

• Fold the square in half, so that *PS* lies exactly on top of *QR*. Crease carefully along *XY*.

Unfold the paper. Mark a point *Z* anywhere on the crease.

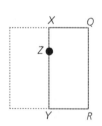

• Fold next along the line through *R* and *Z*; crease and unfold. Then fold along the line through *S* and *Z* to make a third crease line.

2 What is special about the triangle *RSZ*?

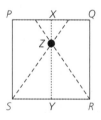

3 a Use folding to find the special point *Z′* on the centre-fold for which the triangle *RSZ′* is a perfect equilateral triangle.

 b Then write out a proof which explains clearly why, when you fold in this way, the triangle *RSZ′* has to be an equilateral triangle.

4 Take the square *PQRS*, with the special point *Z'* (from question **3**) on the centre fold. You know that triangle *RSZ'* is equilateral.

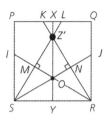

a Fold the side *RS* exactly on top of the line *RZ'*, make a crease along *RM*.

Then fold *SR* exactly on top of *SZ'*; make a crease along *SN*.

b Use a ruler and pencil to mark along all the creases. Your square should look like the one on the right.

c How many marked triangles in your square are congruent to triangle *RSZ'*?

d How many triangles can you find in your square which are congruent to triangle *RYO*?

e Which triangles in your square are *similar* to triangle *RSZ'*? Which are similar to triangle *RYO*?

f Prove that *RO* is twice as long as *OM*.

g Calculate (do not measure) the size of the angles between all the creases. Write them in the appropriate places on your diagram.

5 When you fold the square *PQRS* in half along *XY*, the edge *PS* folds exactly on top of *QR*. So in the labelled square of question **4**, the point *Y* on the bottom edge of the square is the midpoint of the edge *RS*.

The segment *Z'Y* joins the vertex *Z'* of triangle *RSZ'* to the midpoint *Y* of *RS*; *Z'Y* is called a **median** of the triangle *RSZ'*.

The point *M* is the midpoint of *SZ'* (since when *RS* folds exactly on top of *RZ'*, *SM* folds exactly on top of *Z'M*). So the line segment *RM* is also a **median** of the triangle *RSZ'*.

The point *N* is the midpoint of *RZ'* (since when *SR* folds exactly on top of *SZ'*, *RN* folds exactly on top of *Z'N*. So the line segment *SN* is the third **median** of the triangle *RSZ'*.

a What do you notice about the three medians *Z'Y*, *RM*, *SN* in the equilateral triangle *RSZ'*?

b Find out what you can about the three medians of:

an equilateral triangle; an isosceles triangle; a scalene triangle.

6 Cut out the equilateral triangle *RSZ'* with its three fold lines along the medians *Z'Y*, *RM*, *SN*. Explain how to fold the triangle *RSZ'* to obtain a perfect regular hexagon.

19 *Smells, Bells, primes and rhymes*

This activity focuses on:

- exploring ways of counting;
- recognising the same mathematical structure in different settings.

Organisation/Before you start:

- You should have met, and know how to construct, Pascal's triangle.

1. Alys loves the smell of breakfast. She always has egg, bacon and toast. Some days she has all three on the same plate; sometimes all on different plates; and sometimes two items on one plate and the third on another plate.

   a Check that there are 5 different ways of serving these 3 items.

   b If Alys had only 2 items, there would be only 2 different ways of serving them: both items on the same plate, or each item on a separate plate. How many different ways are there of serving a four-item breakfast (for example, egg, bacon, sausage and toast)?

The numbers that come out of question **1** are called **Bell numbers**, named after Eric Temple Bell who studied them in the 1930s. Since there is just one way of serving 1 item, the first Bell number is 1. So the sequence of Bell numbers starts:

$$1, 2, 5, \ldots .$$

These numbers turn up all over the place. But as you found in question **1**, they are not easy to calculate. Imagine trying to keep track of all possible ways of serving a ten-item breakfast! The next question introduces a much simpler way of generating what seem to be the same numbers.

2. **a** Construct a triangle (like **Pascal's triangle**) by writing a 1 at the top.

   Repeat this entry at the start of the second row.
   The next entry in the second row is equal to the **sum** of the number just before it in that row (1) and the number just above in the previous row (1).

1

1   2 (=1+1)

.   .   .

.   .   .   .

b Now you have reached the end of the second row, repeat the last entry (2) at the start of the third row. Then fill in the other two entries in the third row (each being the sum of the previous number in that row and the number just above in the previous row).

c When you reach the end of the third row, repeat the last entry at the start of the fourth row. Then complete the other entries in the fourth row as before.

d Complete the fifth, sixth and seventh rows. Then check the solutions.

e What do you notice about the last entry in each row?

3 Bell numbers also arise when you try to count how many different ways there are to factorise an integer like 30.

a Show that there are 5 different ways to factorise 30 as a 'product' of integers ≥ 2 (you must include the unusual factorisation $30 = 30$, which expresses 30 as a 'product' of just one number).

b In how many different ways can you factorise 210?

c What is the connection between the number of ways of factorising 30 or 210 and the number of different ways of serving Alys's breakfast?

4 One surprising application of Bell numbers is in counting the number of different ways a verse can rhyme. For example, a verse with three lines could have:

• all three lines rhyming (*aaa*), or

• just two rhyming (*aab*, *aba*, *abb*), or

• no two rhyming (*abc*).

a Here we give a typical four line verse (by Tennyson). The rhyming pattern here is *abcb*. List all possible four line rhyming patterns.

b For each of the possible four line rhyming patterns you listed in part **a**, find an example of a four line verse with exactly that rhyming pattern.

c Explain why the number of different rhyming patterns with *n* lines is always equal to the *n*th Bell number.

> *Break, break, break,*
>
> *On thy cold gray*
> *stones, O Sea!*
>
> *And I would that my*
> *tongue could utter*
>
> *The thoughts that*
> *arise in me.*

20 *Folding regular (and irregular) pentagons*

This activity focuses on:

• harder applications of congruence in the context of folding;
• using congruence to analyse interesting geometrical constructions.

Organisation/Before you start:

• You will need a few sheets of plain A4 paper.
• You must complete Challenge 18 before tackling this section.

1   Start with an A4 piece of paper *ABCD*.

Fold the corner *C* exactly on top of the corner *A*, and crease along *XY*.

Do not unfold.

Turn the folded sheet so that *XY* is horizontal.

Fold *DX* exactly on top of *BY* and crease along *AZ;* unfold.

Fold *BY* to lie exactly on the line *AZ*; crease along *RS*.

Fold *DX* to lie exactly on the line *AZ*; crease along *TU*.

The folded pentagon *ARSTU* **looks** suspiciously like a regular pentagon. Is it?

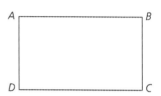

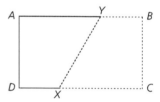

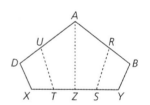

2   Take a strip of paper 1 cm wide and about 15 cm long.

Tie a simple knot in the strip, and gently pull the knot tight.

Press the knot flat as you pull on the two ends.

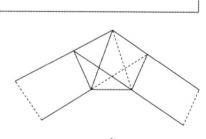

Cut off the two ends of the strip that are not part of the knot.

The flattened knot should form a perfect regular pentagon. However, it is not at all easy to **prove** that it really is a regular pentagon! Try it!

21 Puzzles

CHALLENGES

1 A boat is being winched towards the jetty. If you wind in 1 metre of rope, does the boat move more than 1 metre, exactly 1 metre, or less than 1 metre?

2 Which is larger: the average of the squares of two numbers, or the square of their average?

3 Which fits better, a square peg in a round hole, or a round peg in a square hole?

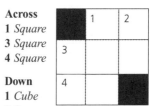

4 A crossnumber is like a crossword puzzle – except that the answers are numbers instead of words. None of the answers starts with the digit 0. How many different solutions are there to this crossnumber?
(You must use logic, not guesswork.)

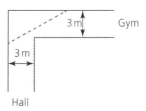

Across
1 *Square*
3 *Square*
4 *Square*

Down
1 *Cube*
2 *Square*
3 *Cube times square*

5 You have to carry a ladder from the Hall to the Gym along this corridor. What is the longest ladder you can get round the corner?

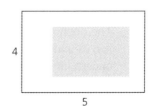

6 **a** Is it true that the sum of three consecutive integers is always divisible by 3? Explain.

   b Is the sum of five consecutive integers always divisible by 5?

   c Is the sum of four consecutive integers always divisible by 4?

7 The shaded rectangle can be turned right round without crossing the sides of the outer rectangle. What is the largest possible value of the perimeter of the shaded rectangle?

Solving and proving

1 It is possible to find an answer without using symbols – but try to spot the hidden assumption!

- The two numbers are in the ratio 2 : 5.

  $\therefore$ The first number has 2 parts; the second number has 5 parts.

  $\therefore$ The two numbers differ by 3 parts.

  $\therefore$ 3 parts is the same as 21.

  $\therefore$ 1 part is the same as 7.

  $\therefore$ The first number (2 parts) must be 14, and the second number 35.

As you read the above solution, you should see that it only works by using the word *part* in place of a symbol. Thus it is usually much simpler to begin by saying:

- Let the two unknown numbers be x and y.

Then $\frac{y}{x} = \frac{5}{2}$

$\therefore \quad y = \frac{5x}{2}.$ (1)

Also $\quad y - x = 21$

$\therefore \qquad y = x + 21.$ (2)

Substituting from equation (2) into equation (1) gives

$x + 21 = \frac{5x}{2}$

$\therefore \ 21 = \frac{3x}{2}$

$\therefore \quad x = \frac{42}{3} = 14$

$\therefore \quad y = 14 + 21 = 35.$ **QED**

The above 'solution' may seem to be complete; but it makes an unjustified assumption – and so fails to find all possible answers.

The opening statement 'Let the two unknown numbers be x and y' leaves undecided which of these two numbers is the biggest. This uncertainty is resolved by equation (2) which effectively says 'Let y be the biggest'.

You are told that 'The two numbers are in the ratio 2 : 5'.
It is natural to think that, since $y > x$, that this says $\frac{y}{x} = \frac{5}{2}$. But this is only correct if x and y are both positive. (For example, $x = {}^-5$ is less than $y = {}^-2$, but $\frac{y}{x} = \frac{5}{2}$.)

The correct solution has to allow for this!

- Let the two unknown numbers be x and y, with $x \le y$.

  Then $x + 21 = y$.   (1)

- The two numbers are in the ratio $2 : 5$.

  Hence either (i) $\frac{y}{x} = \frac{5}{2}$, or (ii) $\frac{x}{y} = \frac{5}{2}$;

  that is, either (i) $2y = 5x$, or (ii) $2x = 5y$.

  If you substitute from equation (1), you get:

  in case (i) $2(x + 21) = 2y = 5x$,

  $\therefore \qquad 42 = 3x$

  $\therefore \qquad x = 14$, so $y = 35$;

  in case (ii) $5(x + 21) = 5y = 2x$,

  $\therefore \qquad 3x = {}^-105$

  $\therefore \qquad x = {}^-35$, so $y = {}^-14$.   **QED**

2 Let the two numbers be x and y, with $x \le y$.

Then $x + 21 = y$.   (1)

The two numbers are in the ratio $2 : 7$.

Hence either (i) $\frac{y}{x} = \frac{7}{2}$, or (ii) $\frac{x}{y} = \frac{7}{2}$;

that is, either (i) $2y = 7x$, or (ii) $2x = 7y$.

If you substitute from equation (1), you get:

in case (i) $2(x + 21) = 2y = 7x$,

$\therefore \qquad 42 = 5x$

$\therefore \qquad x = 8\frac{2}{5}$, so $y = 29\frac{2}{5}$;

in case (ii) $7(x + 21) = 7y = 2x$,

$\therefore \qquad 5x = {}^-147$

$\therefore \qquad x = {}^-29\frac{2}{5}$, so $y = {}^-8\frac{2}{5}$.   **QED**

3 Let the two unknown numbers be x and y, where $\frac{y}{x} = \frac{7}{2}$.

$\therefore \qquad y = \frac{7x}{2}$.   (1)

Also $y.x = 686$

$\therefore \qquad y = \frac{686}{x}$.   (2)

Substituting from equation (2) into equation (1) gives

$\frac{686}{x} = \frac{7x}{2}$

Solving and proving

47

$$\therefore \quad 196 = x^2.$$

At this point it is tempting to say '$x = \sqrt{196}$'. But you have to remember that a quadratic equation always has two roots!

$$\therefore \qquad x = \pm 14.$$

If $x = 14$, then $y = \frac{7x}{2} = 49$.

If $x = {}^-14$, then $y = \frac{7x}{2} = {}^-49$.   **QED**

4 Any odd integer is 1 more than the preceding even integer, so must have the form $2m + 1$, for some even integer $2m$.

If the first odd integer you choose is $2m + 1$, the next will be $2m + 3$. Their product is

$$(2m + 1)\,(2m + 3) = 2m(2m + 3) + 1.\,(2m + 3)$$

$$= (2m)^2 + 6m + (2m + 3)$$

$$= 4m^2 + 8m + 3$$

$$= 4(m^2 + 2m + 1) - 1.$$

Since m is an integer, $m^2 + 2m + 1$ is also an integer, so the product $(2m + 1)\,(2m + 3) = 4(m^2 + 2m + 1) - 1$ is equal to one less than a multiple of 4.   **QED**

Note: The algebra is easier if you write both the odd numbers in terms of the even number between them. Let the first odd number be $2n - 1$. Then the second odd number is $2n + 1$. Their product is then

$$(2n - 1)\,(2n + 1) = (2n)^2 - 1 = 4n^2 - 1.$$ **QED**

5 Write the first odd integer as $2m + 1$.

Then the second and third odd integers are $2m + 3$ and $2m + 5$.

The square of the middle integer is

$$(2m + 3)^2 = (2m)^2 + 2\times(2m)\times3 + 9$$

$$= 4m^2 + 12m + 9.$$

The product of the first and third integers is

$$(2m + 1) \times (2m + 5) = 2m(2m + 5) + 1.\,(2m + 5)$$

$$= 4m^2 + 10m + (2m + 5)$$

$$= (4m^2 + 12m + 9) - 4.$$

So the difference is always equal to 4.   **QED**

Note: The algebra is much easier if you use the fact that the second integer is halfway between the other two. So if you call the middle integer p, then the first integer is $p - 2$, and the third integer is $p + 2$.

The square of the middle integer is p^2.

The product of the first and third integers is $(p - 2)(p + 2) = p^2 - 4$.

So the difference is always equal to 4.   **QED**

6 **a** Let the first of the three consecutive integers be n.

Then the next two integers are $n + 1$ and $n + 2$.

The sum of the last two integers is $(n + 1) + (n + 2) = 2n + 3$.

The sum of the first two integers is $n + (n + 1) = 2n + 1$.

So the difference is always equal to
$(2n + 3) - (2n + 1) = 2$.   **QED**

Note: This is slightly easier if you call the middle integer m.
Then the first and third integers are $m - 1$ and $m + 1$.
The sum of the last two is $m + (m + 1) = 2m + 1$, and the sum of the first two is $(m - 1) + m = 2m - 1$.   **QED**

b Let the middle integer be m.

Then the product of the first and the last is
$(m - 1)(m + 1) = m^2 - 1$, which is 1 less than the square of the middle integer.   **QED**

c Let the middle integer be m.

- Then the sum of all three integers is
$$(m - 1) + m + (m + 1) = 3m.$$

- The product of the last two integers is $m(m + 1) = m^2 + m$.
The product of the first two integers is $(m - 1)m = m^2 - m$.

- The difference between these two products is
$$(m^2 + m) - (m^2 - m) = 2m.$$

And $2m = \frac{2}{3} \times 3m = \frac{2}{3} \times$ (sum of all three integers).   **QED**

7 Let the two-digit integer be 'ab' $= 10a + b$.

Then the reverse 'ba' $= 10b + a$.

Which is the larger of these two integers, depends on which of a and b is largest; so we may as well assume that $a \geq b$. Then
'ab' $\geq$ 'ba'.

Solving and proving

$$\therefore \text{ ‘}ab\text{’} - \text{‘}ba\text{’} = (10a + b) - (10b + a)$$

$$= 9a - 9b$$

$$= 9(a - b).$$

This is clearly a multiple of 9.   **QED**

8 Let the three-digit integer be ‘abc’ $= 100a + 10b + c$.

Then the reverse ‘cba’ $= 100c + 10b + a$.

Which is the larger of these two integers depends on a and c; so we may as well assume that $a \geq c$. Then ‘abc’ $\geq$ ‘cba’.

$$\therefore \text{ ‘}abc\text{’} - \text{‘}cba\text{’} = (100a + 10b + c) - (100c + 10b + a)$$

$$= 99a - 99c$$

$$= 99(a - c).$$

This is clearly a multiple of 99.   **QED**

9 **Claim:** The only possible solutions x, y, z are of the form:
i $0, k, \ ^-k$ for some integer k; **ii** $1, 2, 3$; **iii** $^-1, \ ^-2, \ ^-3$.

Proof: Let the three integers be x, y, and z.

Then $x + y + z = x \cdot y \cdot z$.   (1)

We distinguish four cases:

i one of the three integers $= 0$;

ii one of the three integers $= 1$;

iii one of the three integers $= \ ^-1$;

iv none of the integers $= 0$, or 1, or $^-1$.

i Suppose one of x, y, z equals 0.

$$\therefore \ x \cdot y \cdot z \ = 0.$$

$$\therefore \ x + y + z = 0.$$

$\therefore$ One of the three integers is 0, and the other two must be k and ^-k.

Every triple of the form $0, k, \ ^-k$ satisfies equation (1).

ii Suppose one of the three integers, say z, is equal to 1.

$$\therefore \qquad x + y + 1 \ = x.y.1.$$

$$\therefore \qquad xy - x - y = 1$$

$\therefore\ xy - x - y + 1\ = 2$

$\therefore\ (x - 1)\,(y - 1) = 2.$

So $x - 1$ and $y - 1$ are integers whose product is equal to 2.

$\therefore$ Either **a** one is equal to 1 and the other is equal to 2,

        or **b** one is equal to $^-1$ and the other is equal to $^-2$.

In **a** the three integers must be 1, 2, 3.

In **b**, one of x and y equals 0, so the triple is 1, 0, $^-1$, and the solution has already been listed in case **i** above.

iii Suppose one of the three integers, say z, is equal to $^-1$.

$\therefore\qquad\quad x + y - 1 = x\,.\,y\,.\,^-1$

$\therefore\qquad\quad xy + x + y = 1$

$\therefore\ xy + x + y + 1 = 2$

$\therefore\ (x + 1)\,(y + 1) = 2.$

So $x + 1$ and $y + 1$ are integers whose product is equal to 2.

$\therefore$ Either **a** one is equal to 1 and the other is equal to 2,

        or **b** one is equal to $^-1$ and the other is equal to $^-2$.

In **b** the three integers must be $^-1$, $^-2$, $^-3$.

In **a**, one of x and y equals 0, so the triple is $^-1$, 0, 1, and the solution has already been listed in case **i** above.

iv If any of the three integers x, y, z equals 0, or 1, or $^-1$, then all possible solutions have been found in cases **i**, **ii**, and **iii**.

So you can now assume that $x \geq 2$ or $x \leq {}^-2$; $y \geq 2$ or $y \leq {}^-2$; and $z \geq 2$ or $z \leq {}^-2$.

Let x be the 'numerically largest' of the three integers, and write $|x|$ for the (positive) numerical size of x (so that $|{}^-37| = 37$).

Then the numerical value of the product $x\,.\,y\,.\,z$ is at least 4 times $|x|$ (since the numerical values of y and z are both ≥ 2).

However the numerical value of the sum $x + y + z$ is at most 3 times $|x|$.

So there are no solutions in which each of x, y, z has numerical value ≥ 2. **QED**

10 Let the first integer be n.

Then the four consecutive integers are $n, n + 1, n + 2, n + 3$.

Their product is $n(n + 1)(n + 2)(n + 3)$;

$$n(n + 1)(n + 2)(n + 3) = n(n + 3) \cdot (n + 1)(n + 2)$$
$$= [n^2 + 3n] \cdot [n^2 + 3n + 2].$$

If you let $m = n^2 + 3n + 1$, then this product is $(m - 1)(m + 1) = m^2 - 1$. So the product of four consecutive positive integers is always one less than a perfect square – so cannot itself be a perfect square (since successive squares in the sequence 1, 4, 9, 16, … differ by at least 3).   **QED**

2 *Congruent and similar*

1 b Three pieces of information are enough – as long as you choose carefully (see questions **2** and **3**).

2 The second possibility listed in question **2** is slightly different from the others – and in two ways.

- Firstly, if all you know about a triangle is the size of its three angles, then there are infinitely many different possibilities for the triangle – since any enlargement of a given triangle has the same angles as the original triangle.

This is the only way two triangles can satisfy *AAA*: two triangles *ABC* and *A'B'C'* are **similar** (that is, one is an enlargement of the other) precisely when $\angle CAB = \angle C'A'B'$, $\angle ABC = \angle A'B'C'$, $\angle BCA = \angle B'C'A'$.

- Secondly, the sum of the angles of a triangle is always 180°. So once you know two of the angles in a triangle, you can always work out what the third angle has to be. Hence there is no need to give the size of the third angle.

3 a Any one of *SSS, SAS, ASA, AAS* is enough.

The second bullet in the solution to question **2** explains why *ASA* and *AAS* convey more information than you think: knowing two of the angles in a triangle automatically determines the third angle. So if you tell your friend the size of two of the angles in your triangle, she will always construct a triangle that is similar to your triangle. If you also tell her the length of one particular side, there is no danger that her triangle will be a different size from yours. So *ASA* and *AAS* are enough to ensure that your friend's triangle will be congruent to yours.

This also means that the triple *AAS* is a disguised repeat of the triple *ASA* (since knowing two of the angles in a triangle determines the third angle).

This explains why the standard congruence criteria are

- *SSS*: all three sides

- *SAS*: two sides and the included angle

- *ASA*: two angles and the side between them.

b If $AB = A'B' = 5$, $BC = B'C' = 8$, and $\angle BCA = \angle B'C'A' = 30°$, then the triangles ABC and $A'B'C'$ could be as shown here.

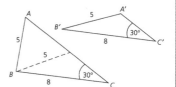

3 *Surprising statistics*

1 **a** 0 would not occur as a first digit.

You might expect each of the other nine digits to occur roughly $\frac{1}{9}$ of the time. But for most sets of data this is not what happens.

b I got the following results.

Digit	0	1	2	3	4	5	6	7	8	9
Number of occurrences as first digit	0	99	84	53	49	35	26	18	22	19
Relative frequency ≈	0	0.24	0.21	0.13	0.12	0.09	0.06	0.04	0.05	0.05

2 This should become clear in question **4**.

4 Here is the relative frequency chart for the data in the table above.

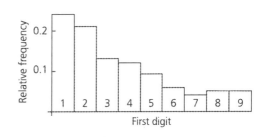

5 **a** If there were exactly 99 houses in the street, each non-zero digit would occur 11 times as a first digit, so each relative frequency would be $\frac{11}{99} = \frac{1}{9}$.

b If there were exactly 999 houses in the street, each non-zero digit would occur 111 times as a first digit, so each relative frequency would be $\frac{111}{999} = \frac{1}{9}$.

c If there were exactly 110 houses in the street, the digit 1 would occur as a first digit 11 + 11 times as a first digit, its relative frequency would be $\frac{22}{110} = \frac{2}{11}$;

every other digit would still only occur 11 times as a first digit, so their relative frequency drops to $\frac{11}{110} = \frac{1}{10}$.

If there were exactly 222 houses in the street, the digit 1 would occur as a first digit 100 + 11 times, so would have relative frequency $\frac{111}{222} = \frac{1}{2} = 0.5$;

the digit 2 would occur 23 + 11 times as a first digit, so would have relative frequency $\frac{34}{222} = \frac{17}{111} \approx 0.153$;

every other digit would still only occur 11 times as a first digit, so their relative frequency drops to $\frac{11}{222} \approx 0.05$.

6 At first you might expect the new set of figures to behave very differently. The original list contained lots of numbers with first digit 1, and when you multiply by 2 they **all** change to numbers with first digit different from 1! But you already know that first impressions can be misleading.

For example, although the list

10,	12,	14,	15,	17,	19,	22,	25,	28,	32,
35,	38,	43,	47,	53,	57,	61,	73,	85,	97,

contains only twenty numbers, it more or less fits Benford's distribution: 1 occurs as first digit 6 times, 2 occurs as first digit 3 times, 3 occurs as first digit 3 times, and so on.

When you double each of the numbers you may get a surprise:

20,	24,	28,	30,	34,	38,	44,	50,	56,	64,
70,	76,	86,	94,	106,	114,	122,	146,	170,	194.

1 occurs as first digit 6 times, 2 occurs 3 times, 3 occurs 3 times, and so on. The pattern is not quite the same, but it is much closer than you would expect.

The precise mathematical reason for this is beyond the scope of this book. On a practical level it may help to realise that, if Benford's Law is valid for lists of measurements, then it cannot depend on the **unit** used to do the measuring: and measuring weights in pounds instead of kilograms is a bit like doubling all the answers!

Note: Benford's law states that if you randomly select a number from a large table of statistical data, then the probability that the first digit is d is equal to $\log_{10}\left(1 + \frac{1}{d}\right)$. Use your calculator to find

$$\log_{10}\left(1 + \frac{1}{1}\right) = \log_{10}2, \quad \log_{10}\left(1 + \frac{1}{2}\right) = \log_{10}\left(\frac{3}{2}\right), \quad \log_{10}\left(\frac{4}{3}\right),$$

$$\log_{10}\left(\frac{5}{4}\right), \quad \log_{10}\left(\frac{6}{5}\right), \quad \log_{10}\left(\frac{7}{6}\right), \quad \log_{10}\left(\frac{8}{7}\right), \quad \log_{10}\left(\frac{9}{8}\right).$$

7 In 1995, *The Wall Street Journal* (the North American equivalent of the *Financial Times*) published an article under the headline: 'He's got their number. Scholar uses math to find financial fraud'. The methods now used by major accounting firms are more sophisticated but they are based on the principle that when people invent numbers, the digit patterns of the numbers they invent cause the data to appear unnatural, and this can often be identified using certain computer checks.

 Right-triangle numbers

1 a Three (3, 4, 5; 5, 12, 13; 20, 21, 29).

   b One such triple is: 7, 24, 25.

     Start with 3 and 4. Then $25 = 3^2 + 4^2$; $24 = 2 \times 3 \times 4$; and $7 = 4^2 - 3^2$.

2 a • Take any two starting integers p, q.

     • Square and add to get $p^2 + q^2$: this is the hypotenuse.

     • Take the difference of the two squares: $p^2 - q^2$. Then work out twice the product of the two starting integers: $2pq$. These are the two shorter sides.

     • The triple: $p^2 - q^2$, $2pq$, $p^2 + q^2$ is a Pythagorean triple.

3 **a** $z = (a^2 + b^2)^2 = a^4 + 2a^2b^2 + b^4$

$$= a^4 - 2a^2b^2 + b^4 + 4a^2b^2$$

$$= (a^2 - b^2)^2 + (2ab)^2 = y^2 + x^2.$$

  b Start with the two integers $a = 2$, $b = 1$;
we get the triple $z = 5 = 2^2 + 1^2$, $x = 4 = 2 \times 2 \times 1$,
and $y = 3 = 2^2 - 1^2$.
This gives the first row of the table on the right.

a, b	z	x	y
2, 1	5	4	3
3, 1	10	6	8
4, 1	17	8	15
5, 1	26	10	24
6, 1	37	12	35
3, 2	13	12	5
4, 2	20	16	12
5, 2	29	20	21
6, 2	40	24	32
4, 3	25	24	7
5, 3	34	30	16
6, 3	45	36	27
5, 4	41	40	9

4 The integer 15 cannot be written as a sum of two squares.

∴ The triple 9, 12, 15 is never produced by the method Mariotte describes.

Note: The related triples 3, 4, 5 and 27, 36, 45 are produced by the method Mariotte describes. Diophantos understood exactly why this happens, and – 1400 years before Mariotte – wrote his own book much more carefully than Mariotte. It was this book by Diophantos that inspired Fermat around 1630 to make the conjecture which became known as **Fermat's Last Theorem**, and which was eventually proved in 1994 by Andrew Wiles.

5 **b** Let $x = 2m + 1$,

Then $x^2 = 4m^2 + 4m + 1$

$$= (2m^2 + 2m) + (2m^2 + 2m + 1)$$

Let $y = 2m^2 + 2m$, $z = 2m^2 + 2m + 1$.

Then $z^2 = (2m^2 + 2m + 1)^2$

$$= (2m^2 + 2m)^2 + 2(2m^2 + 2m) + 1$$

$$= (2m^2 + 2m)^2 + 4m^2 + 4m + 1$$
$$= (2m^2 + 2m)^2 + (2m + 1)^2$$
$$= y^2 + x^2. \quad \textbf{QED}$$

 # 5 *Percentages*

1 a $10 + 20 + 30 + 40 = 100$.

$\therefore$ 100% of pupils score A, B, C or D.

$\therefore$ No pupils score E or worse.

b $10 + 15 + 20 + 25 = 70$.

$\therefore$ 70% of the class score D or better.

6 pupils score E or worse, so these 6 pupils make up 30% of the class.

Let the number of pupils in the class be x.

$\therefore 6 = (30\% \text{ of } x) = \frac{30}{100} \cdot x$

$\therefore x = 20$.

2 To calculate 10% of £30 you must work out $\frac{10}{100} \times £30$.

To calculate 30% of £10 you must work out $\frac{30}{100} \times £10$.

So the two answers are the same.

3 I receive a discount of 10% on each item;

$\therefore$ Each squeegee costs $\frac{10}{100} \times £20 = £2$ less than usual.

$\therefore$ Each squeegee costs £18.

$\therefore$ Three squeegees cost $3 \times £18 = £54$.

Note 1: A 10% discount means that I pay 90% of the original price. Three squeegees should cost $3 \times £20 = £60$;

$\therefore$ I actually pay $\frac{90}{100} \times £60 = £54$.

Note 2: '10% off each item' does **not** mean 30% off three items!

4 Suppose you currently receive £P per week.
Then £P represents 100% of your current pocket money.

COMMENTS & SOLUTIONS

A 30% increase will lead to you receiving 130% of this; that is, $£\frac{130}{100} \times P$.

A further 40% increase the following year would lead to your receiving 140% of this new total; that is,
$$£\frac{140}{100} \times \left(\frac{130}{100} \times P\right) = £\frac{182}{100} \times P.$$

This is the same as a 40% rise followed by a 30% rise
$$£\frac{130}{100} \times \left(\frac{140}{100} \times P\right) = £\frac{182}{100} \times P - \text{namely an 82\% increase overall.}$$

Note 1: Since the overall rise is the same in each case, you might choose the second option, since this gives you more money in the first year.

Note 2: Though the combined increase is the same in each case, it is definitely not equal to $(30 + 40)\%$.

5 30% of 40% of 50% of £100 $= \frac{30}{100} \times \frac{40}{100} \times \frac{50}{100} \times (£100) = £6$.

6 Cost price = £39.

Marked up price $= \frac{4}{3} \times (£39) = £52$.

$\therefore$ Sale price $= \frac{75}{100} \times (£52) = \frac{3}{4} \times \left(\frac{4}{3} \times (£39)\right) = £39$.

7 a Suppose mock-turtle soup initially costs £M per serving.

An increase of 50% would mean that the new price is $£\frac{150}{100} \times M = £\frac{3}{2} \times M$.

A further increase of $33\frac{1}{3}\%$ would mean that the new price is
$£\frac{4}{3} \times \left(\frac{3}{2} \times M\right) = £(2M)$; that is, a 100% increase overall.

b A drop of 50% on day 3 would mean that the new price is $£\frac{50}{100} \times (2M) = £M$.

A drop of $33\frac{1}{3}\%$ on day 4 would mean that the final price is $£\frac{2}{3} \times M$.

So the overall percentage change in the price over the four days would be a $33\frac{1}{3}\%$ decrease.

c A drop of $33\frac{1}{3}\%$ on day 3 would mean that the new price is $£\left(\frac{2}{3} \times 2M\right) = £\frac{4}{3}M$.

A further drop of 50% on day 4 would give a final price of $£\frac{2}{3}M$.

So the overall percentage change in the price over the four days would again be a $33\frac{1}{3}\%$ decrease.

The reason why the answers to parts **b** and **c** are the same is that percentage changes are multiplicative.

In part **b** the original price of £M per serving changes to
$£\left(\frac{2}{3} \times \frac{1}{2} \times \frac{4}{3} \times \frac{3}{2} \times M\right)$;

In part **c** the original price of £M per serving changes to
$£\left(\frac{1}{2} \times \frac{2}{3} \times \frac{4}{3} \times \frac{3}{2} \times M\right)$.

8 Price of meal plus 17.5% VAT $= \frac{117.5}{100} \times £12.50$.

Price of meal plus VAT plus 12.5% service charge

$$= \frac{112.5}{100} \times \frac{117.5}{100} \times £12.50. \qquad (1)$$

Price of meal plus 12.5% service charge $= \frac{112.5}{100} \times £12.50$.

Price of meal plus service charge plus 17.5% VAT

$$= \frac{117.5}{100} \times \frac{112.5}{100} \times £12.50. \qquad (2)$$

The two calculations (1) and (2) clearly give the same answer.

9 **a** Let there be N sheep in the original flock.

The shepherd first loses 60% of her flock – that is $\frac{60}{100} \times N$ sheep.

She then finds 60% of these;

∴ 40% of these $\frac{60}{100} \times N$ sheep are still missing.

∴ $\frac{40}{100} \times \frac{60}{100} \times N = \frac{24}{100} \times N$ sheep are still missing.

That is, 24% of the flock is still missing.

b Let there be N sheep in the original flock.

The shepherd loses $\frac{3}{4}$ of his flock – that is, $\frac{75}{100} \times N$ sheep.

He then finds all except 10% of the missing sheep;

∴ 10% of $\frac{75}{100} \times N$ sheep are still missing – that is,
$\frac{10}{100} \times \frac{75}{100} \times N = \frac{7.5}{100} \times N$ sheep.

∴ 7.5% of the flock is still missing.

10 **a** Let the junior pitch have length l and width w.

Then the senior pitch has length $\frac{110}{100}l$ and width $\frac{110}{100}w$.

∴ Area of senior pitch $= \frac{110}{100}l \times \frac{110}{100}w$

$$= \left(\frac{110}{100} \times \frac{110}{100}\right) \times (l \times w).$$

$$= \left(\frac{110}{100} \times \frac{110}{100}\right) \times (\text{area of junior pitch}).$$

$$= \left(\frac{121}{100} \times \text{area of junior pitch}\right).$$

So the senior pitch requires 21% more seed than the junior pitch.

b Let the senior pitch have length L and width W.

Then $L = \frac{110}{100}l$ and $W = \frac{110}{100}w$; so $l = \frac{100}{110}L \approx \frac{91}{100}L$ and $w = \frac{100}{110}W \approx \frac{91}{100}W$.

$$\therefore \text{ Area of junior pitch} = l \times w = \frac{100}{110}L \times \frac{100}{110}W$$

$$= \left(\frac{100}{110} \times \frac{100}{110}\right)(L \times W)$$

$$= \left(\frac{100}{110} \times \frac{100}{110}\right) \times (\text{area of senior pitch})$$

$$\approx \frac{82.6}{100} \times (\text{area of senior pitch}).$$

So the junior pitch requires 17.4% less seed than the senior pitch.

(More simply: $S = \frac{121}{100} \times J$, so $J = \frac{100}{121} \times S \approx \frac{82.6}{100} \times S$)

c 10% more.

11 a 20%

The radius r increases by 20% to $\frac{120}{100}r$.

$\therefore$ The circumference increases from $2\pi r$ to
$2\pi\left(\frac{120}{100}r\right) = \frac{120}{100} \times (2\pi r)$.

b 44%

The radius r increases by 20% to $\frac{120}{100}r$.

$\therefore$ The area increases from πr^2 to $\pi\left(\frac{120}{100}r\right)^2 = \frac{144}{100} \times (\pi r^2)$.

12 a The radius r increases to $2r$.

$\therefore$ The area increase from πr^2 to $\pi(2r)^2 = 4\pi r^2$.

$\therefore$ The area increases by 300%.

b Suppose the area doubles, and that the radius increases from r to R.

The area increases from πr^2 to $\pi R^2 = 2(\pi r^2)$.

$\therefore R^2 = 2r^2$

$\therefore R = (\sqrt{2})r \approx 1.414r$

$\therefore r$ increases by roughly 41.4%.

13 Let the cuboid have length l, width w, height h.

$\therefore$ Volume of original cuboid $= l \times w \times h$.

Each length is increased by 50%, so the new cuboid has length $\frac{3}{2}l$, width $\frac{3}{2}w$, and height $\frac{3}{2}h$.

a Volume of new cuboid $= \left(\frac{3}{2}l \times \frac{3}{2}w \times \frac{3}{2}h\right)$

$= \frac{27}{8}\ (l \times w \times h) = \frac{337.5}{100} \times$ (volume of original cuboid).

$\therefore$ Volume increases by 237.5%.

b Area of l by w face of original cuboid $= lw$.

Area of corresponding face of new cuboid
$= \left(\frac{3}{2}l\right) \times \left(\frac{3}{2}w\right) = \frac{9}{4}(lw) = \frac{225}{100}(lw)$.

$\therefore$ Surface area of l by w face increases by 125%.

Similarly for every other face.

$\therefore$ Total surface area increases by 125%.

14 Let the original cuboid have length l, width w and height h.

$\therefore$ Volume $= l \times w \times h$.

Each length is increased by $P\%$.

$\therefore$ Each length is multiplied by $x = \frac{100 + P}{100}$.

$\therefore$ Volume $= lx \times wx \times hx = (lwh)\,x^3$.

Volume is doubled, so $x^3 = 2$.

$\therefore x = \sqrt[3]{2} \approx 1.2599$.

$\therefore$ An increase of almost 26% is needed in each length to double the volume.

15 Let the circles $\mathcal{C}$ and $\mathcal{C}'$ have radius r.

The radius of $\mathcal{C}$ is doubled from r to $2r$; then the area increases from πr^2 to $4\pi r^2$.

The area is then halved from $4\pi r^2$ to $2\pi r^2$; so the radius decreases from $2r$ to $(\sqrt{2})r$ (since $\pi(\sqrt{2} \cdot r)^2 = 2\pi r^2$).
$\therefore$ Circle $\mathcal{C}_1$ has radius $(\sqrt{2})r$.

The area of $\mathcal{C}'$ is first doubled from πr^2 to $2\pi r^2$; so the radius changes from r to $(\sqrt{2})r$ (since $\pi(\sqrt{2} \cdot r)^2 = 2\pi r^2$).
The radius is then halved from $(\sqrt{2})r$ to $\frac{(\sqrt{2})r}{2}$;
$\therefore$ Circle $\mathcal{C}'_1$ has radius $\frac{(\sqrt{2})r}{2}$.

∴ Since $\frac{(\sqrt{2})r}{2} < r < (\sqrt{2})r$,

circle $\mathcal{C}'_1$ is the smallest and circle $\mathcal{C}_1$ is the largest of $\mathcal{C}, \mathcal{C}_1, \mathcal{C}'_1$.

16 Each year the price increases by 5%, so the price is multiplied by $\frac{105}{100} = \frac{21}{20}$.

In fifteen years, the price is multiplied by $\frac{21}{20}$ fifteen times, so changes from £50 000, to $\left(\frac{21}{20}\right)^{15} \times £50\,000 \approx £103\,946.41$, an increase of almost 108%.

6 *Divisibility rules*

1 b Start with the two-digit number '*ab*'.

The rule uses the expression $5b + a$, whereas the original number '*ab*' $= 10a + b$.

To get an expression with $10a$ in it, we multiply $5b + a$ by 10:

$$10(5b + a) = 50b + 10a = (10a + b) + 49b.$$

The last term $49b$ on the right-hand side is always a multiple of 7.

∴ 7 divides the whole right-hand side $(10a + b) + 49b$ precisely when 7 divides $(10a + b) = $ '*ab*'.

The **highest common factor** of 7 and 10 is 1: $hcf\,(7, 10) = 1$.

∴ 7 divides the left-hand side $10(5b + a)$ precisely when 7 divides $5b + a$.

So 7 divides '*ab*' $= 10a + b$ precisely when 7 divides $5b + a$.

QED

c Use exactly the same rule, but replace 'the tens digit' by 'the number of tens'.

Claim: Take any three-digit integer '*abc*'.
Multiply the units digit by 5.
Add the number of tens to the answer.
Then the original number is divisible by 7
precisely when the final answer is divisible by 7.

Proof: '*abc*' $= 100a + 10b + c = 10(10a + b) + c$, so the number of tens is precisely $(10a + b)$.

364		
4×5	$= 20$	
$20 + 36$	$= 56$	
56	$= 7 \times 8$	
∴ 7 goes into 364		

You have to show that

$100a + 10b + c$ is divisible by 7 precisely when $5c + (10a + b)$ is divisible by 7.

$$10 \,[5c + (10a + b)] = 100a + 10b + 50c$$
$$= (100a + 10b + c) + 49c.$$

Now $49c$ is always divisible by 7. So 7 divides the right-hand side precisely when 7 divides the original number 'abc' = $(100a + 10b + c)$.

And, since $hcf\,(7, 10) = 1$, 7 divides the left-hand side precisely when 7 divides the bracket $[5c + (10a + b)]$.

So 7 divides 'abc' precisely when 7 divides $5c + (10a + b)$.

QED

2 a Start with the two-digit number 'ab'.

The rule uses the expression $4b + a$, whereas the original number 'ab' = $10a + b$.

To get an expression with $10a$ in it we multiply $4b + a$ by 10:

$$10\,(4b + a) = 40b + 10a = (10a + b) + 39b.$$

The last term $39b$ on the right-hand side is always a multiple of 13.

∴ 13 divides the whole right-hand side $(10a + b) + 39b$ precisely when 13 divides $(10a + b)$ = 'ab'.

Since $hcf\,(13, 10) = 1$
∴ 13 divides the left-hand side $10(4b + a)$ precisely when 13 divides $4b + a$.

So 13 divides 'ab' = $10a + b$ precisely when 13 divides $4b + a$.

QED

b Use exactly the same rule, but replace 'the tens digit' by 'the number of tens'. The same proof works as in question **1c**.

3 a The original number 'ab' = $10a + b$.

You want a rule which uses an expression of the form $m{\times}b + a$.

To get an expression with $10a$ in it we multiply $m{\times}b + a$ by 10:

$$10\,(m{\times}b + a) = (10m)b + 10a = 10a + b + (10m - 1)b.$$

So you must choose m so that $10m - 1$ is always a multiple of 19. Clearly $m = 2$ works.

Divisibility rules

Test for divisibility by 19: Given any two-digit integer, multiply the units digit by 2, then add the tens digit.
- If the answer is a multiple of 19, so is the original number.
- If the answer is not a multiple of 19, neither is the original number.

Proof: $10(2b + a) = 20b + 10a = 10a + b + 19b$.

∴ 19 divides the right-hand side precisely when 19 divides $10a + b =$ 'ab'. Moreover, $hcf(19, 10) = 1$.

∴ 19 divides the left-hand side precisely when 19 divides $2b + a$.

So 19 divides 'ab' $= 10a + b$ precisely when 19 divides $2b + a$.
QED

b This extends to three-digit integers.

4 The original number 'ab' $= 10a + b$.

You want a rule which uses an expression of the form $m \times b + a$.

To get an expression with $10a$ in it we multiply by 10:

$$10(m \times b + a) = (10m)b + 10a = 10a + b + (10m - 1)b.$$

So you must choose m so that $10m - 1$ is always a multiple of 23. Clearly $m = 7$ works.

Test for divisibility by 23: Given any two-digit integer, multiply the units digit by 7, then add the tens digit.
- If the answer is a multiple of 23, so is the original number.
- If the answer is not a multiple of 23, neither is the original number.

Proof: $10(7b + a) = 70b + 10a = 10a + b + 69b$.

∴ 23 divides the right-hand side precisely when 23 divides $10a + b =$ 'ab'. Moreover, $hcf(23, 10) = 1$.

∴ 23 divides the left-hand side precisely when 23 divides $7b + a$.

So 23 divides 'ab' $= 10a + b$ precisely when 23 divides $7b + a$.
QED

The rule extends to three-digit integers.

7 *Fraction Puzzles*

1 a $1\frac{1}{2} \times \frac{2}{3} = \frac{3}{2} \times \frac{2}{3} = 1$; $2\frac{2}{3} \times \frac{3}{4} = \frac{8}{3} \times \frac{3}{4} = 2$; $3\frac{3}{4} \times \frac{4}{5} = \frac{15}{4} \times \frac{4}{5} = 3$.

b $10\frac{10}{11} \times \frac{11}{12} = \frac{120}{11} \times \frac{11}{12} = 10$.

c $49\frac{49}{50} \times \frac{50}{51} = \frac{2499}{50} \times \frac{50}{51} = 49$.

d Claim: $\left(n + \frac{n}{n+1}\right) \times \frac{n+1}{n+2} = n$.

Proof: $\left(n + \frac{n}{n+1}\right) = \frac{n(n+1)}{n+1} + \frac{n}{n+1}$

$$= \frac{n^2 + 2n}{n+1} = \frac{n(n+2)}{n+1}.$$

$\therefore \left(n + \frac{n}{n+1}\right) \times \frac{n+1}{n+2} = \frac{n(n+2)}{n+1} \times \frac{n+1}{n+2}$

$$= n. \quad \textbf{QED}$$

2 a Suppose you choose 7.

The reciprocal is $\frac{1}{7}$.

Subtracting from 1 gives $1 - \frac{1}{7} = \frac{6}{7}$.

The reciprocal of $\frac{6}{7}$ is $\frac{7}{6}$.

Subtracting from 1 gives $1 - \frac{7}{6} = -\frac{1}{6}$.

The reciprocal of $-\frac{1}{6}$ is $^{-}6$.

Subtracting from 1 gives $1 - (^{-}6) = 7$.

So you are back at the start and repeat the sequence: $7, \frac{6}{7}, -\frac{1}{6}, \ldots$.

b Suppose you start with the integer n.

The reciprocal is $\frac{1}{n}$.

Subtracting from 1 gives $1 - \frac{1}{n} = \frac{n-1}{n}$.

The reciprocal of $\frac{n-1}{n}$ is $\frac{n}{n-1}$.

Subtracting from 1 gives $1 - \frac{n}{n-1} = \frac{n-1}{n-1} - \frac{n}{n-1} = \frac{-1}{n-1}$.

The reciprocal of $\frac{-1}{n-1}$ is $-(n-1)$.

Subtracting from 1 gives $1 - (-(n-1)) = 1 + (n-1) = n$.

So you are back at the start and repeat the sequence: $n, \frac{n-1}{n}, \frac{-1}{n-1}, \ldots$.

3 You are told that a and b are positive integers and that $\frac{a}{b}$ is a proper fraction. So $0 < a < b$.

To compare fractions you have to put them over a common denominator:

a $\frac{a}{b} = \frac{a(b+1)}{b(b+1)}$ and $\frac{a+1}{b+1} = \frac{b(a+1)}{b(b+1)}$.

Now $\qquad\quad a < b$

$\therefore \qquad ab + a < ab + b$

$\therefore \qquad a(b+1) < b(a+1)$

$\therefore \qquad \frac{a(b+1)}{b(b+1)} < \frac{b(a+1)}{b(b+1)}$

$\therefore \quad \frac{a(b+1)}{b(b+1)} = \frac{a}{b} < \frac{a+1}{b+1} = \frac{b(a+1)}{b(b+1)}.$ **QED**

b To get from $\frac{a-1}{b-1}$ to $\frac{a}{b}$, you add 1 to numerator and denominator – just as in part **a**. So, provided $a-1$ and $b-1$ are positive (that is, provided $a \geq 2$ and $b \geq 2$)

$\therefore \frac{a-1}{b-1} < \frac{a}{b}.$

c You know from part **a** that $\frac{a}{b} < \frac{a+1}{b+1}$.

And again from part **a** that $\frac{a+1}{b+1} < \frac{a+2}{b+2}$.

$\therefore \frac{a}{b} < \frac{a+2}{b+2}.$

d **Claim:** If $0 < a < b$, and $n > 0$, then $\frac{a}{b} < \frac{a+n}{b+n}$.

Proof: Put both fractions over a common denominator:

$\frac{a}{b} = \frac{a(b+n)}{b(b+n)}$, and $\frac{a+n}{b+n} = \frac{b(a+n)}{b(b+n)}$.

Now $a < b$ and $n > 0$.

$\therefore \qquad\quad an < bn$

$\therefore \quad ab + an < ab + bn$

$\therefore \quad a(b+n) < b(a+n).$

Divide both sides by $b(b+n)$:

$\therefore \frac{a(b+n)}{b(b+n)} = \frac{a}{b} < \frac{a+n}{b+n} = \frac{b(a+n)}{b(b+n)}.$ **QED**

4 **a** **Claim:** $\frac{a}{b} < \frac{a+c}{b+d} < \frac{c}{d}$.

b **Proof:** Notice first that $0 < \frac{a}{b} < \frac{c}{d}$.

Write both fractions with common denominator bd:

$\therefore 0 < \frac{ad}{bd} < \frac{bc}{bd}$

$\therefore ad < bc. \qquad\qquad (1)$

To prove the **Claim** you have to prove

i $\frac{a}{b} < \frac{a+c}{b+d}$, and **ii** $\frac{a+c}{b+d} < \frac{c}{d}$.

i Add ab to both sides of inequality (1):

$\therefore\ ab + ad < ab + bc$

$\therefore\ a(b + d) < b(a + c).$ (2)

Divide both sides of inequality (2) by $b(b + d)$:

$\therefore\ \frac{a}{b} < \frac{a+c}{b+d}.$

ii Add cd to both sides of the inequality (1):

$\therefore\ ad + cd < bc + cd$

$\therefore\ (a + c)d < (b + d)c.$ (3)

Divide both sides of inequality (3) by $d(b + d)$:

$\therefore\ \frac{a+c}{b+d} < \frac{c}{d}.$ **QED**

 8 *Behold Pythagoras!*

Each of the diagrams illustrates a completely general proof of Pythagoras' theorem – provided one looks at it in the right way, and then uses the diagram to write out the full proof.

1 This illustrates the classical proof (Euclid's *Elements*, Book 1, Proposition 47) for the particular case when the right-angled triangle ABC is a 3-4-5 triangle.

Let the perpendicular from C to BA meet BA at X.

Then $\triangle XCB$ is similar to $\triangle CAB$.

$\therefore\ BX : BC = BC : BA$

$\therefore\ BX : 4 = 4 : 5$

$\therefore\ BX = \frac{16}{5},\ XA = \frac{9}{5}.$

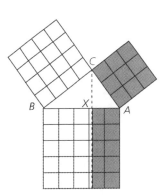

So the perpendicular from C to BA cuts the square on BA into two rectangles: one 5 by $\frac{16}{5}$, the other 5 by $\frac{9}{5}$.

Notice that: $5 \times \frac{16}{5} = 4^2$, and $5 \times \frac{9}{5} = 3^2$.

$\therefore\ \text{Area(square on } BA) = 5^2$

$$= 5 \times \frac{16}{5} + 5 \times \frac{9}{5}$$

$$= 4^2 + 3^2$$

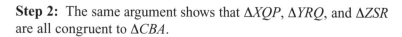= area(square on CB) + area(square on CA).

2 Suppose you are given a right-angled triangle ABC with a right angle at C.

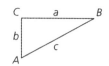

The sides have lengths $CA = b$ and $CB = a$, and the hypotenuse has length $AB = c$.

Construct a square $WXYZ$ of side length $a + b$.

Go round the square, marking points P, Q, R, S, one on each side, each distance a from the previous corner.

Join up the four marked points P, Q, R, S.

The quadrilateral $PQRS$ looks like a square, but you have to **prove** that it really is a square.

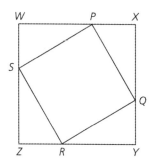

Step 1: $WXYZ$ was constructed to be a square.

∴ $\angle PWS$ is a right angle.

Also $WP = a$ (by construction)

and $WS = WZ - SZ$

$$= (a + b) - a$$

$$= b \text{ (by construction)}.$$

∴ $\triangle CBA$ and $\triangle WPS$ are congruent (by the SAS congruence criterion, since $CB = WP = a$, $\angle BCA = \angle PWS = 90°$, and $CA = WS = b$).

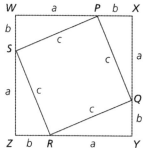

Step 2: The same argument shows that $\triangle XQP$, $\triangle YRQ$, and $\triangle ZSR$ are all congruent to $\triangle CBA$.

∴ $SP = PQ = QR = RS = c$

and $\angle SPQ = \angle PQR = \angle QRS = \angle RSP$.

Hence the angles of quadrilateral $PQRS$ are all equal.

Since they must have sum 360°, it follows that

∴ $\angle SPQ = \angle PQR = \angle QRS = \angle RSP = 90°$.

The quadrilateral $PQRS$ has four equal sides and four right angles, so it is a square.

Step 3: Now work out the area of the square $WXYZ$ in two different ways.

Area (square $WXYZ$) = $(a + b)^2 = a^2 + 2ab + b^2$.

Area (square $WXYZ$) = area (square $PQRS$) + area (four corner triangles)

$$= c^2 + 4 \times \left(\tfrac{1}{2}ab\right).$$

$$\therefore \qquad a^2 + 2ab + b^2 = c^2 + 4 \times \left(\tfrac{1}{2}ab\right)$$

$$\therefore \qquad a^2 + b^2 = c^2. \quad \textbf{QED}$$

3 Given a right-angled triangle CBA with $AB = c$, $BC = a$, $CA = b$.

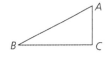

Four copies of this right-angled triangle can be fitted together to make the square with side equal to the hypotenuse – *with a hole in the middle*.

The large square clearly has area c^2.

The diagram on page 16 shows that the four triangles and the shaded square hole in the middle can be rearranged to make:
a square of side a, together with a square of side b.

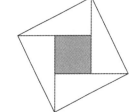

So this rearrangement has area $a^2 + b^2$.

$$\therefore \ a^2 + b^2 = c^2. \quad \textbf{QED}$$

4 Given a right-angled triangle CBA with $AB = c$, $BC = a$, $CA = b$.

Four copies of this right-angled triangle can be fitted together to make the square with side of length $(a + b)$, the sum of the two shorter sides, and *with a hole in the middle*.

The large square clearly has area $(a + b)^2$.

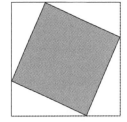

The four triangles can be rearranged in a different way to make two corners of a square of exactly the same size, namely of area $(a + b)^2$; what is left is:
an a by a square (bottom left) and a b by b square (top right).

The areas of the two large squares are equal.

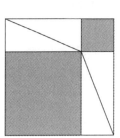

And the four triangles in each figure have the same total area.

$\therefore$ Shaded area in the first figure = shaded area in second figure.

$$\therefore \ c^2 = a^2 + b^2. \quad \textbf{QED}$$

5 The large square $ABCD$ is a c by c square.

The triangles $\triangle ABX$ and $\triangle DCY$ are **similar** right-angled triangles (since corresponding sides are parallel), with $AB = DC$. So $\triangle ABX$ and $\triangle DCY$ are **congruent**.

Similarly $\triangle DAW$ and $\triangle CBZ$ are congruent.

Also $DA = AB$ (since $ABCD$ is a square).

And

$\angle DAW + \angle XAB = \angle DAB = 90°.$

$\therefore \ \angle DAW = 90° - \angle XAB = \angle ABX.$

$\therefore \ \angle ADW = 90° - \angle DAW = \angle XAB.$

$\therefore \ \triangle DAW$ and $\triangle ABX$ are congruent (by the *ASA* congruence criterion).

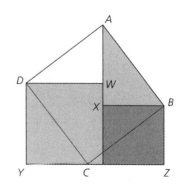

So the four right-angled triangles – two at the top and two at the bottom of the diagram – are all congruent to each other.

Hence the diagram can be looked at in two different ways,

On the one hand,

total area = area (c by c square $ABCD$) + area (two lower triangles)

$\qquad = c^2 +$ area (two lower triangles).

On the other hand,

total area = area (two upper triangles) + area (two lower squares)

$\qquad =$ area (two upper triangles) + $(a^2 + b^2)$.

$\therefore \ c^2 = a^2 + b^2.$ **QED**

6 Given a right-angled triangle CBA with $AB = c$, $BC = a$, $CA = b$.

Two copies of this right-angled triangle and half of a square of side c have combined area of

$$2 \times \left(\tfrac{1}{2}ab\right) + \tfrac{1}{2}c^2,$$

and can be fitted together, to make the trapezium with parallel sides of lengths a and b, and with perpendicular distance between them equal to $a + b$.

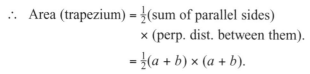

$\therefore \ $ Area (trapezium) $= \tfrac{1}{2}$(sum of parallel sides)

$\qquad\qquad\qquad \times$ (perp. dist. between them).

$\qquad\qquad = \tfrac{1}{2}(a + b) \times (a + b).$

$\therefore \ \tfrac{1}{2}(a + b) \times (a + b) = 2 \times \left(\tfrac{1}{2}ab\right) + \tfrac{1}{2}c^2$

$\therefore \qquad a^2 + 2ab + b^2 = 2ab + c^2$

$\therefore \qquad\qquad a^2 + b^2 = c^2.$ **QED**

COMMENTS & SOLUTIONS

Note: This proof of Pythagoras' theorem was published by James Abram Garfield, 20th President of the United States of America.

7   Given a right-angled triangle CBA with $AB = c$, $BC = a$, $CA = b$.

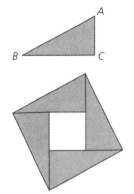

Four copies of this right-angled triangle can be fitted together to make the square with side equal to the hypotenuse – *with a hole in the middle.*

The large square clearly has area c^2.

The hole is a square with side of length $a - b$.

$\therefore\ \ c^2$ = area of square hole + area of 4 triangles

$$= (a - b)^2 + 4 \times \left(\tfrac{1}{2}ab\right)$$

$$= (a^2 - 2ab + b^2) + 2ab$$

$$= a^2 + b^2. \quad \textbf{QED}$$

1 a

Number of beads	Number of necklaces
1	2
2	3
3	4
4	6
5	8

b

c | 6 | 14 (*13?*) |

Two necklaces are counted as being the same if one picture can be obtained by rotating the other one. So the three-bead necklace

    black-black-white is the same as **black-white-black**.

When we get to 6 beads, there are 14 necklaces if you count as before and only allow rotations.

However, two of these 14 necklaces are indistinguishable if you are allowed to reflect the picture (or turn the necklace over).

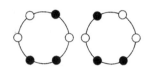

2 a

Total	Number of ways
1	1
2	2
3	3
4	5
5	7

b

6	11
7	15

3 a

Total	Number of ways
1	1
2	2
3	3
4	5
5	8

b

6	13
7	21

c Claim: Each term in the sequence is obtained by adding the two previous terms.

Proof: Let S_n denote the number of ways of climbing a staircase with n steps.

The last move is either **i** short (1 step) or **ii** long (2 steps).

Case i If the last move is short, then the previous moves were used to climb a staircase of $n - 1$ steps.

There are S_{n-1} ways of climbing a staircase of $n - 1$ steps.

So there are S_{n-1} ways of climbing n steps with the final move being short (1 step).

Case ii If the last move is long, then the previous moves were used to climb the first $n - 2$ steps.

There are S_{n-2} ways of climbing a staircase of $n - 2$ steps.

So there are S_{n-2} ways of climbing n steps with the final move being long (2 steps).

So there are $S_{n-1} + S_{n-2}$ ways to climb a staircase of n steps.

$\therefore\ S_n = S_{n-1} + S_{n-2}$. **QED**

4 This is exactly the same as question **3** (since $1 + 1 + 2 + 2 + 2 + 2$ is another way of saying 'short-short-long-long-long-long').

So the number of ways of writing 10 is the tenth number in the sequence

$$1, 2, 3, 5, 8, 13, 21, 34, 55, \mathbf{89}, 144, \ldots$$

from question **3**.

5 **a**

Total	Number of ways
1p	1
2p	2
3p	2
4p	3
5p	3
6p	4
7p	4

b Claim: If N is even, the number of ways to make Np is $\frac{N+2}{2}$.

If N is odd, the number of ways to make Np is $\frac{N+1}{2}$.

Proof: The key idea is to concentrate on how many 2p stamps you use. If you use no 2p stamps, the whole Np has to be made up using 1p stamps. If you use exactly one 2p stamp, the remaining $(N - 2)$p has to be made up using 1p stamps. And so on.

- If N is even, the number of 2p stamps you use can be

$$0, \text{ or } 1, \text{ or } 2, \ldots, \text{ or } \tfrac{N}{2}.$$

$\therefore$ The number of ways to make Np is exactly $\frac{N}{2} + 1 = \frac{N+2}{2}$.

- If N is odd, the number of 2p stamps you can use is equal to

$$0, \text{ or } 1, \text{ or } 2, \ldots , \text{ or } \frac{N-1}{2}.$$

$\therefore$ The number of ways to make Np is exactly $\frac{N-1}{2} + 1 = \frac{N+1}{2}$

QED

6 a

Total	Number of ways
1p	1
2p	2
3p	3
4p	4
5p	5
6p	7
7p	8

b There is no easy answer here – so don't worry if you tried and failed. The following comments show how the same idea as in the solution of question **5b** can be used predict the number of ways of making Np in this question – provided you break it up into different cases.

Suppose $N = 3k$ is a multiple of 3. Then the number of 3p stamps used is

$$0, \text{ or } 1, \text{ or } 2, \ldots , \text{ or } \frac{N}{3} = k.$$

- If 0 3p stamps are used, you have to make Np using only 1p and 2p stamps – and question **5b** tells you how many ways there are to do this: if N is even there are $\frac{N+2}{2}$ ways, and if N is odd there are $\frac{N+1}{2}$ ways.

- If 1 3p stamp is used, you have to make the remaining $(N-3)$p using only 1p and 2p stamps; again question **5b** tells you how many ways there are to do this:
if N is even, then $N-3$ is odd so there are $\frac{(N-3)+1}{2}$ ways, and
if N is odd, then $N-3$ is even and there are $\frac{(N-3)+2}{2}$ ways.

Continuing like this shows that:

if N is an even multiple of 3, then the number of ways to make Np

$$= \frac{N+2}{2} + \frac{(N-3)+1}{2} + \frac{(N-6)+2}{2} + \frac{(N-9)+1}{2} + \ldots + \frac{3+1}{2} + \frac{0+2}{2},$$

while if N is an odd multiple of 3, then the number of ways to make Np

$$= \tfrac{N+1}{2} + \tfrac{(N-3)+2}{2} + \tfrac{(N-6)+1}{2} + \tfrac{(N-9)+2}{2} + \dots + \tfrac{3+1}{2} + \tfrac{0+2}{2}.$$

Similar formulas hold when N is not a multiple of 3. In question **5b** there were two slightly different answers; here you get six slightly different answers.

7 When a sequence starts 1, 2, 4, 8, … , you may expect it to continue … , 16, 32, 64, 128, … .

Whether it does in fact continue in this way will depend on where the sequence comes from.

a The terms go 1, 2, 4, 8, … . This is a genuine sequence of powers of 2.

Claim: There are exactly 2^n ways to write the positive integer n as a sum of one or more positive integers.

Proof: Let W_n denote the number of ways of writing n as a sum of positive integers. Then $W_1 = 1$, $W_2 = 2$, $W_3 = 4$.

Any way of writing n as a sum of positive integers either

i begins with '1 + …', or

ii begins with an integer $k \geq 2$.

Case i Any expression that begins '1 + …' is obtained by tacking '1 + ' onto the front of an expression for $n - 1$.

So there are exactly W_{n-1} ways of writing n which begin '1 + …'.

Case ii Any expression which begins 'k + …' with $k \geq 2$ is obtained by increasing by 1 the first term of an expression

$$(k - 1) + \dots.$$

for $n - 1$.

So there are exactly W_{n-1} ways of writing n which begin 'k + …' with $k \geq 2$.

∴ There are exactly $W_{n-1} + W_{n-1} = 2W_{n-1}$ ways to write n as a sum of positive integers.

∴ $W_n = 2W_{n-1}$.

This guarantees that each term is twice the preceding term.

Since the sequence starts with $W_1 = 2^0$, it follows that $W_n = 2^{n-1}$. **QED**

b This sequence starts out looking like powers of 2.

But the sequence is in fact very different.

With n points on the circle, the largest possible number of regions is equal to

$$\frac{n(n-1)}{24}\left(n^2 - 5n + 18\right) + 1.$$

Points	Number of regions
1	1
2	2
3	4
4	8
5	16
6	31

c If 3 seats are contested, the possible outcomes are

    BBB;   *BBL*, *BLB*, *LBB*;   ***BLL***, ***LBL***, ***LLB***;   *LLL*;

so there are exactly 4 **odd** results for the Big-Enders.

If 4 seats are contested, the odd outcomes for the Big-Enders are:

    BBBL, *BBLB*, *BLBB*, *LBBB*;   *BLLL*, *LBLL*, *LLBL*, *LLLB*;

so there are exactly 8 **odd** results.

Claim: If n seats are contested, the number of odd results for Big-Enders is exactly 2^{n-1}.

Proof: Let the number of **odd** results for Big-Enders when n seats are contested be O_n, and let the number of **even** results be E_n.

$$\therefore \quad O_1 = 1 = E_1. \qquad (1)$$

In general $O_n + E_n$ is equal to the total number of possible results. The first seat can be won by either B or L – so there are 2 possible outcomes. The same is true for each of the n seats. So the total number of possible outcomes with n seats is $2 \times 2 \times 2 \times \ldots \times 2 = 2^n$.

$$\therefore \quad O_n + E_n = 2^n. \qquad (2)$$

In any odd result with n seats, the first seat in the list of winners is either **i** a B, or **ii** an L.

Case i If the first seat is a B, then the list of winners is obtained by tacking a B on the front of an **even** result for $n - 1$ seats.

$$\therefore \quad \text{There are exactly } E_{n-1} \text{ odd results of type i.}$$

Case ii If the first seat is an L, then the list of winners is obtained by tacking a L on the front of an **odd** result for $n - 1$ seats.

∴ There are exactly O_{n-1} **odd** results of type **ii**.

∴ $O_n = E_{n-1} + O_{n-1}.$ (3)

This is all you need.

You already know from (1) that $O_1 = 1 = 2^0$ and $E_1 = 1 = 2^0$.

∴ $O_2 = E_1 + O_1 = 2^0 + 2^0 = 2^1$ (from (3)).

From (2) you then know that $E_2 = 2^2 - O_2$
$$= 2^2 - 2^1 = 2^1.$$

So now you know that $O_2 = E_2 = 2^1$.

∴ $O_3 = E_2 + O_2 = 2^1 + 2^1 = 2^2$ (from (3)).

From (2) you then know that $E_3 = 2^3 - O_3$
$$= 2^3 - 2^2 = 2^2.$$

So now you know that $O_3 = E_3 = 2^2$.

Continuing in this way shows that $O_n = E_n = 2^{n-1}$.   **QED**

10 *Onwards and upwards*

Your scattergrams should indicate the trend suggested in the title of the section! Except for minor blips, the general trend is for distances to increase steadily with time. In some events (such as the men's pole vault and the women's javelin) the winning performance does appear to improve **linearly**. Can performance continue to improve linearly?

The exceptional blips may take the form of:

a unexpected improvements caused by some new style, such as the Fosbury flop in the High Jump, or due to the location of the games making longer throws or jumps easier;

b apparent drops in performance caused by a new rule, such as a change in the design or weight of the discus.

You might like to look at the corresponding results for the women's events. What interesting questions does this data suggest?

Year	High jump	Long jump	Discus
1928	159		3962
1932	165		4058
1936	160		4763
1948	168	569	4192
1952	167	624	5142
1956	176	635	5369
1960	185	637	5510
1964	190	676	5727
1968	182	682	5828
1972	192	678	6662
1976	193	672	6900
1980	197	706	6996
1984	202	696	6536
1988	203	740	7230
1992	202	714	7006

A useful reference: *Olympic Games Records*, by Stan Greenberg (published by Guinness).

11 *Very* small

1 $0.1\,\text{mm} = 10^{-1}\,\text{mm} = 1 \times 10^{-4}\,\text{m}$.

2 The answer depends on your weight.

Suppose you weigh 50 kg; 65% of this is water.

∴ Weight of water in your body $\approx 0.65 \times 50$ kg

$$= 32.5 \text{ kg}.$$

Each molecule of water weighs roughly 3×10^{-23} kg.

∴ Each kg of water contains $\frac{1}{3 \times 10^{-23}} = \frac{1}{3} \times 10^{23}$ molecules.

∴ A body weighing 50 kg contains roughly

$$32.5 \times \tfrac{1}{3} \times 10^{23} \approx 1 \times 10^{24} \text{ water molecules.}$$

3 The answer depends on the type of paper – but the most widely used ordinary paper is '80 gsm' (which means '80 grams per square metre').

If you measure a pile of 100 sheets of 80 gsm paper, you will find that it is almost exactly 1 cm high.

∴ 1 sheet has thickness 0.01 cm = 0.1 mm, or 1×10^{-1} mm.

4 a A pollen grain is about 0.01 mm = 1×10^{-2} mm in diameter.

   b A virus is about 0.0001 mm = 1×10^{-4} mm in diameter.

   c A hydrogen atom is about 1×10^{-7} mm in diameter.

6 1 teaspoon ≈ 5 ml

$$= 5 \text{ cubic cm.}$$

1 acre ≈ 0.405 hectares

$$= 0.405 \times 100 \times 100 \text{ square metres}$$

$$= 0.405 \times 10\,000 \times 10\,000 \text{ square cm}$$

$$= 4.05 \times 10^{7} \text{cm}^{2}.$$

5 cm^{3} of oil spreads over an area of 4.05×10^{7} cm^{2}, and volume of oil layer = thickness of oil layer × area.

∴ 5 = thickness × 4.05×10^{7}

∴ Thickness of oil layer $= \frac{5}{4.05 \times 10^{7}}$ cm ≈ 1.2×10^{-7} cm.

12 *Fractions and decimals*

1 a $0.4 = \tfrac{4}{10}$ is one obvious answer.

   Two others might be the fractions corresponding to 0.35 and 0.45, namely $\tfrac{35}{100} = \tfrac{7}{20}$ and $\tfrac{45}{100} = \tfrac{9}{20}$.

2 You could do this directly – knowing that $\frac{1}{2} = \frac{4}{8}$ and that $\frac{3}{4} = \frac{6}{8}$, so that $\frac{5}{8}$ for example must lie between $\frac{1}{2}$ and $\frac{3}{4}$. Similarly $\frac{9}{16}$ and $\frac{11}{16}$ also lie between $\frac{1}{2} = \frac{8}{16}$ and $\frac{3}{4} = \frac{12}{16}$.

Alternatively, you could use the fact that $\frac{1}{2} = 0.5$, $\frac{3}{4} = 0.75$; then choose three decimals which lie between 0.5 and 0.75, say 0.55, 0.6 and 0.7; finally change these back into fractions to get $0.55 = \frac{11}{20}$, $0.6 = \frac{6}{10} = \frac{3}{5}$, $0.7 = \frac{7}{10}$.

3 $\frac{7}{10}$ is closer to $\frac{2}{3}$.

Proof: $\frac{3}{4} - \frac{7}{10} = \frac{15}{20} - \frac{14}{20} = \frac{1}{20}$

$\qquad \frac{7}{10} - \frac{2}{3} = \frac{21}{30} - \frac{20}{30} = \frac{1}{30}$.

4 **a** $0.25 = \frac{25}{100} = \frac{1}{4}$

   b $0.24 = \frac{24}{100} = \frac{6}{25}$

   c $1.25 = \frac{125}{100} = \frac{5}{4}$

   d $0.625 = \frac{625}{1000} = \frac{5}{8}$

   e $0.5625 = \frac{5625}{10\,000} = \frac{9}{16}$

   f $1.234 = \frac{1234}{1000} = \frac{617}{500}$

5 **a** $4\overline{)1\,.\,0\,0}\;^{0\,.\,2\,5}$; $\frac{1}{4} = \frac{1 \times 25}{4 \times 25} = \frac{25}{100} = 0.25$

   b $\frac{7}{50} = 0.14$

   c $\frac{9}{8} = 1.125$

   d $\frac{7}{20} = 0.35$

   e $\frac{13}{25} = 0.52$

   f $\frac{5}{32} = 0.156\,25$

6 **a** $3\overline{)1\,.\,0\,0\,0\,0\,0\,\ldots}\;^{0\,.\,3\,3\,3\,3\,3\,\ldots}$; $\frac{1}{3} = 0.\dot{3}$

   b $\frac{1}{6} = 0.1\dot{6}$

   c $\frac{1}{7} = 0.\dot{1}42\,85\dot{7}$

   d $\frac{1}{9} = 0.\dot{1}$

e $\frac{1}{11} = 0.\dot{0}\dot{9}$

f $\frac{3}{7} = 0.\dot{4}28\,57\dot{1}$

7 a $0.\dot{5} = \frac{5}{9}$

    Proof: Let $x = 0.\dot{5}$.

    $\therefore \qquad 10x = 5.\dot{5}$

    $\therefore \; 10x - x = 5.555\,555\ldots \text{(forever)} - 0.555\,555\ldots \text{(forever)}$
                $= 5$

    $\therefore \qquad\quad 9x = 5.$ **QED**

b $0.8\dot{3} = \frac{5}{6}$

c $0.\dot{1}\dot{5} = \frac{15}{99} = \frac{5}{33}$

d $0.\dot{0}\dot{9} = \frac{9}{99} = \frac{1}{11}$

e $0.1\dot{2}\dot{3} = \frac{1}{10} + \frac{23}{990} = \frac{122}{990} = \frac{61}{495}$

f $0.\dot{1}2\dot{3} = \frac{123}{999} = \frac{41}{333}$

8 a In the word ∗e∗ai∗∗e∗∗, only the vowels have been included; the consonants r, m, n, d, r, s have been omitted.

b $11\overline{\smash{)}2\,.\,{}^{2}0\,{}^{9}0\,{}^{2}0\,0\,0\,0\ldots}$ $0\,.\,1\,8\,1\,.\,.\,.$

c The third remainder is the same as the first.

    $\therefore$ The division at the third stage repeats the first stage, with the same remainder.

    $\therefore$ The fourth remainder will be the same as the second, and so on.

    $\therefore$ The first and second steps are repeated over and over again.

9 a $7\overline{\smash{)}3\,.\,{}^{3}0\,{}^{2}0\,{}^{6}0\,{}^{4}0\,{}^{5}0\,{}^{1}0\,{}^{3}0}\ldots$ $0\,.\,4\,.\,.\,.\,.\,.$

b $13\overline{\smash{)}4\,.\,{}^{4}0\,{}^{1}0\,{}^{10}0\,{}^{9}0\,{}^{12}0\,{}^{3}0\,{}^{4}0}\ldots$ $0\,.\,3\,.\,.\,.\,.\,.$

c $19\overline{\smash{)}1\,.\,{}^{1}0\,{}^{10}0\,{}^{5}0\,{}^{12}0\,{}^{6}0\,{}^{3}0{}^{11}0{}^{15}0{}^{17}0{}^{18}0\,{}^{9}0{}^{14}0\,{}^{7}0{}^{13}0{}^{16}0\,{}^{8}0\,{}^{4}0\,{}^{2}0\,{}^{1}0}\ldots$ $0\,.\,0\,5\,.\,.\,.\,.\,.\,.\,.\,.\,.\,.\,.$

10 Suppose you do the division to find the decimal for the fraction $\frac{1}{23}$.

At each step of the division process, the remainder must be less than 23. So there are exactly 23 possible remainders – namely 0, 1, 2, ... , 22.

In this case the remainder 0 never occurs; but if it did occur, then the decimal would terminate.

Suppose the decimal does not terminate.

Then there are exactly 23 – 1 possible remainders: namely 1, 2, ... , 22. So by the time you get to the 23rd step, one of these remainders must have appeared for the second time, and the whole division process will have started to recur.

In general suppose you do the division to find the decimal for the fraction $\frac{p}{q}$.

At each step of the division process, the remainder must be less than q. So there are exactly q possible remainders: namely 0, 1, 2, ... , $q - 1$.

If the remainder 0 occurs, then the decimal terminates.

Suppose the decimal does not terminate.
Then there are exactly $q - 1$ possible remainders: namely 1, 2, ... , $q - 1$. So by the time you get to the qth step, one of these remainders must have appeared for the second time, and the whole division process will have started to recur.

13 *Pigeons and pigeon-holes*

1 a If Sandra chooses ≤ 3 socks, they could all be of different colours.

∴ Sandra must choose at least 4 socks to obtain a pair.

Suppose Sandra's socks are red, white and blue.

Consider three pigeon-holes – one for each colour.

If Sandra chooses at least 4 socks, then some pigeon-hole receives at least two socks;
∴ Sandra is guaranteed a pair of socks of the same colour.

COMMENTS & SOLUTIONS

b Suppose Sandra's socks are of n different colours.

If Sandra chooses $\leq n$ socks, they could all be of different colours.

$\therefore$ Sandra must choose at least $n + 1$ socks to obtain a pair.

Consider n pigeon-holes – one for each colour.

If Sandra chooses at least $n + 1$ socks, then some pigeon-hole receives at least two socks;
$\therefore$ Sandra is guaranteed a pair of socks of the same colour.

2 The solutions in this problem use a variation of the pigeon-hole principle.

  a Let the chosen integers be the 'pigeons'.

Imagine two pigeon-holes – one for integers < 50, and the other for integers ≥ 50.

Put each integer in the appropriate pigeon-hole.

If neither integer is ≥ 50, **then the second pigeon-hole remains empty**.

$\therefore$ The two integers both go in the < 50 pigeon-hole.

$\therefore$ The two integers have sum ≤ 98 – which is false. **QED**

  b Suppose five integers have sum 100.

Let the five chosen integers be the 'pigeons'.

Imagine two pigeon-holes – one for integers < 20, and the other for integers ≥ 20.

Put each integer in the appropriate pigeon-hole.

If none of the five integers is ≥ 20, **then the second pigeon-hole remains empty**.

$\therefore$ All five integers go in the < 20 pigeon-hole.

$\therefore$ The five integers have sum $\leq 5 \times 19 = 95$ – which is false.

$$\textbf{QED}$$

3 Imagine one million pigeon-holes (or boxes) – one for each positive integer $n \leq 1\,000\,000$.

Let the heads in London be the pigeons.

If a head has n hairs, put that head in the box marked n.

Since there are far more than 1 million heads in London, two of them must land up in the same box.

4 Suppose there are n pupils at Hogwarts School. Then each pupil could have 0, or 1, or 2, . . ., or $n - 1$ friends among the other pupils.

Label the pigeon-holes with the numbers 0, 1, 2, … , $n - 1$.

And let the n pupils be the pigeons.

Each pupil knows a definite number of other pupils – either 0, or 1, or 2, … , or $n - 1$. Put each pupil in the pigeon-hole for his number of friends.

a You are told that 'each pupil knows at least one other pupil'.

So the pigeon-hole labelled '0' remains empty.

∴ The n pupils are allocated to only $n - 1$ pigeon-holes.

∴ Some pigeon-hole receives at least 2 pigeons.

So there must be at least two pupils with the same number of friends.

b Suppose you are **not** told that 'each pupil knows at least one other pupil'.

Then there would seem to be n pigeons (the n pupils) and n pigeon-holes (marked 0, 1, 2, . . ., $n - 1$). So you might think that there could be one pigeon in each pigeon-hole.

However, if some pupil has $n - 1$ friends, then he knows all the other pupils, so no pupil can be friendless: that is, the pigeon-holes marked 0 and $n - 1$ cannot **both** be occupied.

Hence there are n pigeons and at most $n - 1$ occupied pigeon-holes.

∴ Some pigeon-hole receives at least two pigeons.

So there still exist at least two pupils with the same number of friends.

5 a Yes.

This is exactly like question **4b**.

Each team plays each of the $n - 1$ other teams once.

So each team may draw 0, or 1, or 2, . . ., or $n - 1$ games during the season.

Let the n teams be the pigeons, and let the pigeon-holes be marked 0, 1, 2, . . ., $n - 1$.

If a team ends the season with m draws, then it goes in the pigeon-hole marked m.

If some team, say team A, draws $n - 1$ games, then this team draws every game. So every team draws at least one game (since every team draws the game against A).

∴ Either the pigeon-hole marked 0 is empty, or the pigeon-hole marked $n - 1$ is empty.

∴ There are n teams and at most $n - 1$ non-empty pigeon-holes.

So some pigeon-hole has to contain at least two pigeons: that is, at the end of the season there must always be two teams who have drawn exactly the same number of games.

b No.

Each team plays each of the other teams twice.

∴ Each team plays $2(n - 1) = 2n - 2$ games;

∴ Each team could draw 0, or 1, or 2, . . ., or $2n - 2$ games.

So the possible number of draws for each team ($2n - 2$) is usually much larger than the number of teams (n), so it seems unlikely that, at the end of the season, there will always be two teams with the same number of draws.

When $n = 2$, the two teams only play each other, so both teams draw the same number of games. But when $n \geq 3$ there is no reason why any two teams should draw the same number of games. For example, if there are three teams – A, B, C – then there are six games: AB, BA, BC, CB, CA, AC. If the only draws were the games AC, BC, CB, then A would draw once, B would draw twice, and C would draw three times.

6 a Suppose you choose

3,   9,   14,   21,   26,   35,   42,   59,   63,   76.

Then $14 + 63 = 77$ and $35 + 42 = 77$.

Also $3 + 9 + 14 = 26$, and $26 = 26$.

b Let the three integers be a, b, c.

Then there are $7 = 2^3 - 1$ possible subsets you could choose:

$\{a\}$, $\{b\}$, $\{c\}$, $\{a,b\}$, $\{b,c\}$, $\{c,a\}$, $\{a,b,c\}$.

Let these seven subsets be the pigeons.

Let the pigeon-holes be labelled by the possible remainders you can get when dividing by 7: that is, 0, 1, 2, 3, 4, 5, 6.

For each subset, decide which pigeon-hole it goes in like this:

- add up the numbers in the subset

- divide the total by 7 and take the remainder

- put the subset in the pigeon-hole labelled by that remainder.

You want to prove that

- either some subset has a sum which is a multiple of 7,

- or there are two subsets whose sums differ by a multiple of 7.

Suppose that no subset has sum equal to a multiple of 7.

∴ The pigeon-hole labelled 0 remains empty.

∴ The 7 subsets are allocated to just 6 pigeon-holes.

∴ Some pigeon-hole recieves at least two subsets – and these two subsets have sums which differ by a multiple of 7.   **QED**

c With four integers there are $15 = 2^4 - 1$ possible subsets.

d With 10 integers, there are $2^{10} - 1 = 1023$ possible subsets.

Claim: Any set of n integers has $2^n - 1$ non-empty subsets.

Proof: (The proof is very similar to the proof on page 76 of equation (2) in the solution to question **7c** in Challenge 9.)

Let the n integers be $N_1, N_2, N_3, \dots , N_n$.

When choosing a subset, you must first decide whether to include N_1, then whether to include N_2, then whether to include N_3, and so on.

For N_1 you have 2 choices (include or omit); for N_2 you also have 2 choices; similarly for each of the n integers N_i you have 2 choices. So the number of possible ways of choosing a subset is equal to

$2 \times 2 \times 2 \times \dots \times 2 = 2^n$.

This method of counting includes the 'empty subset' (where each integer N_i is omitted). So the number of non-empty subsets is $2^n - 1$.   **QED**

Each of the ten integers is < 100;

∴ The sum of any subset is $< 10 \times 100 = 1000$.

COMMENTS & SOLUTIONS

Take 1000 pigeon-holes, labelled 1, 2, 3, ... , 1000.

Let the 1023 subsets be the pigeons.

Put each subset in the pigeon-hole which is labelled by the sum of the integers in the subset.

Since 1023 > 1000, some pigeon-hole receives at least two different subsets.

∴ There must exist two different subsets which have the same sum.   **QED**

14 *A short history of* π

1 π is the Greek letter *pi* (like our *p*); δ is the Greek letter *delta* (like our *d*). π was short for *perimeter* and δ was short for *diameter*; so $\frac{\pi}{\delta}$ stood for the ratio *perimeter : diameter*.

2 a The diameter was 10 cubits; the circumference was 30 cubits.

b The ratio *circumference : diameter* is taken as 3.

3 a Ahmes multiplies 8 khet by 8 khet and gets 64 setat.

∴ 1 setat must be the same as 1 **square khet**.

b The diameter of the field is about 450 metres. This would make it a huge field, so it seems unlikely that the problem was about a real field. It looks as though the number 9 was chosen to make the first step of the solution – 'Subtract $\frac{1}{9}$ of the diameter' – as easy as possible.

c $\pi r^2 = \pi \left(\frac{D}{2}\right)^2 = \pi \cdot \frac{D^2}{4}$

$$= \pi \frac{81}{4} \approx 63.617\,25 \dots .$$

d $\left(\frac{8}{9}D\right)^2$ is being used in place of $\left(\frac{\pi}{4}\right)D^2$.

∴ π is taken to be $4 \times \left(\frac{8}{9}\right)^2 \approx 3.1605$.

4 a Perimeter of hexagon = 6*r*.

Circumference of circle = 2π*r*.

Perimeter of hexagon : circumference of circle = $\frac{57}{60} + \frac{36}{60^2}$.

Note: The number was written like this because the Babylonians wrote their numbers using **base 60**, not base 10.

$$\therefore \quad \frac{6r}{2\pi r} = \frac{57}{60} + \frac{36}{60^2}.$$

$$\therefore \quad \frac{3}{\pi} = \frac{57}{60} + \frac{36}{60^2}.$$

b $\quad \dfrac{1}{\pi} = \dfrac{19}{60} + \dfrac{12}{60^2}.$

$$\therefore \quad \pi = \frac{60^2}{60 \cdot 19 + 12} = \frac{300}{5 \cdot 19 + 1} = \frac{300}{96} = \frac{25}{8} = 3\tfrac{1}{8} = 3.125.$$

5 a Claim 1: The regular 12-gon inscribed in a circle of radius r has sides of length $s = r.\sqrt{(2 - \sqrt{3})}$.

Proof: Let A, B, C, D, E, F be six points equally spaced around a circle of radius r.

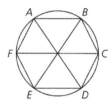

$\therefore$ $ABCDEF$ is a regular hexagon inscribed in the circle.

$\therefore$ $r = AB = BC = CD = DE = EF = FA$.

Let the perpendicular bisectors of the sides AB, BC, CD,

DE, EF, FA meet the circle at U, V, W, X, Y, Z.

Then $AUBVCWDXEYFZ$ is a regular 12-gon inscribed in the same circle of radius r. Moreover, if O is the centre of the circle and if M is the midpoint of AB, then $AM = \frac{r}{2}$.

Apply Pythagoras to triangle OMA to get

$$OM^2 = OA^2 - AM^2 = r^2 - \left(\tfrac{r}{2}\right)^2 = \tfrac{3}{4}r^2$$

$$\therefore \quad OM = \tfrac{r\sqrt{3}}{2}.$$

$$\therefore \quad UM = UO - OM$$

$$= r - \tfrac{r\sqrt{3}}{2} = r\left(1 - \tfrac{\sqrt{3}}{2}\right).$$

Now triangle AMU is a right-angled triangle, so, by Pythagoras we have:

$$AU^2 = AM^2 + UM^2$$

$$\therefore \quad AU^2 = \left(\tfrac{r}{2}\right)^2 + r^2\left(1 - \tfrac{\sqrt{3}}{2}\right)^2$$

$$= \tfrac{r^2}{4} + r^2\left(\tfrac{7}{4} - \sqrt{3}\right) = r^2(2 - \sqrt{3}).$$

$$\therefore \quad AU = r.\sqrt{(2 - \sqrt{3})} = s. \quad \textbf{QED}$$

Claim 2: $6.\sqrt{(2 - \sqrt{3})} < \pi$.

Proof: The perimeter of the 12-gon $= 12s = 12r.\sqrt{(2 - \sqrt{3})}$.

The circumference of the circle = $2\pi r$.

Since the 12-gon is inscribed in the circle, we have $12s < 2\pi r$.

$\therefore\ 12r.\sqrt{(2-\sqrt{3})} < 2\pi r$

$\therefore\ 6.\sqrt{(2-\sqrt{3})} < \pi$.   **QED**

b The regular hexagon is made up of six equilateral triangles, each with height r.

$\therefore$ Each equilateral triangle has side length $t = \frac{2r}{\sqrt{3}}$.

$\therefore\ 6 . \left(\frac{2r}{\sqrt{3}}\right) > \pi D = 2\pi r$

$\therefore\ \pi < 2\sqrt{3} \approx 3.4641$.

Note: You will notice that the overestimate $2\sqrt{3} \approx 3.4641$ is a rather poor approximation to π. This reflects the fact that the circumscribed regular 6-gon is a poor approximation to the circle. Similarly, the underestimate $6.\sqrt{(2-\sqrt{3})} \approx 3.105$ (from part **a**) is not a very good approximation to π.
This may help to explain why Archimedes had to go as far as an inscribed regular 96-gon.

6 $\frac{355}{113}$ = 3.141 592 92 … . This is very close to the exact value of π (π = 3.141 592 653 589 793 …). We have no idea how Zu Chongzhi discovered this approximate value.

 15 *Interesting! But why does it work?*

1 a The question with a 1p and a 2p coin is very easy, because there are only two possible 'answers':

- If the 1p coin is in the left hand, then the answer is $3\times2 + 2\times1 = 8$.

- If the 1p coin is in the right hand, the answer is $3\times1 + 2\times2 = 7$.

So if the answer is even (8), the 1p coin must be in the left hand; and if the answer is odd (7), the 1p coin must be in the right hand.

b The same principle works whenever one coin is odd and the other is even.

- If the odd coin is in the left hand, then the answer is 3×(even) + 2×(odd), which is even.

- If the odd coin is in the right hand, the answer is 3×(odd) + 2×(even), which is odd.

So if the answer is even, the odd coin must be in the left hand.

2   The final score is equal to the sum of the five numbers:

- the three numbers which are visible at the end, plus

- the score a on the dice that was picked up, plus

- the number b on the bottom of the dice that was picked up.

The magician can see the three numbers which are visible at the end – so can quickly work out their sum S.

The numbers a and b are on opposite faces of the dice, and opposite faces have sum 7: $\therefore a + b = 7$.

$\therefore$ The final score is $S + 7$.

3   Let the scores on the three dice by x, y, z.

Then your partner works out

first $2x$;

then $2x + 5$;

then $5(2x + 5) = 10x + 25$;

then $10x + 25 + y = 10x + y + 25$;

then $10(10x + y + 25) = 100x + 10y + 250$;

then $100x + 10y + 250 + z = (100x + 10y + z) + 250$.

All you have to do is to subtract 250 and read off the three digits x, y, z of the answer.

4   Let the unknown page number be p, the line number be l, and the word number be w.

Your partner works out

first $2p$;

then $10 \times 2p = 20p$;

then $5 \times 20p = 100p$;

then $100p + l$;

then $100p + l + 5$;

then $100 \times (100p + l + 5) = 10\,000p + 100l + 500$;

then $10\,000p + 100l + 500 + w = 10\,000p + 100l + w + 500$;

then $10\,000p + 100l + w + 500 + 611 = 10\,000p + 100l + w + 1111$.

All you have to do is to subtract 1111 and read off

- the number formed by the last two digits (that is, the digits in the tens and units columns) = w;

- the number formed by the next two digits (that is, the digits in the thousands and hundreds columns) = l;

- the number formed by the next three digits (that is, the digits in the millions, the hundred thousands and the ten thousands columns) = p.

16 *Area puzzles*

1 Suppose you construct a quadrilateral $ABCD$ with one side, say AD, of length 12 and the other three sides of lengths 3, 4, 5.

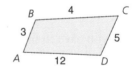

Then the distance from A to D along AD would be equal to the distance from A to D via B and C – which is possible only if B and C lie on AD. So the quadrilateral has area 0.

2 **Claim:** In all three squares the shaded region has the same area.

Proof: Let each of the squares have sides of length $2r$.

i The semicircles in the first square have radius r. So the two semicircles have total area $\frac{1}{2}\pi r^2 + \frac{1}{2}\pi r^2 = \pi r^2$.

ii The quarter circle in the second square has radius $2r$, so has total area $\frac{1}{4}(\pi\,[2r]^2) = \pi r^2$.

iii Each of the quarter circles in the third square has radius r, so the total shaded area is $4 \times \frac{1}{4}(\pi r^2) = \pi r^2$. **QED**

3 Let the large circle have area X and the small circle have area Y.

Let the overlap have area Z.

When the circles are arranged as shown,

$$A = X - Z \text{ and } B = Y - Z.$$

$$\therefore \ A - B = (X - Z) - (Y - Z)$$

$$= X - Y.$$

So the two shaded regions always differ by the fixed amount $X - Y$.

Note: When the small circle is completely inside the large circle then $B = 0$, $A = X - Y$ (and $Z = Y$), and it is obvious that the difference $A - B = X - Y$.

4 Claim: Area of bow-tie $= \frac{17}{8}x$.

Proof: Triangle XAB has area $\frac{1}{2} \times 3 \times h$ (using AB as base of length 3). Triangle $XA'B'$ has area $\frac{1}{2} \times 5 \times h'$ (using $A'B'$ as base of length 5).

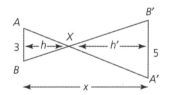

$\therefore$ Area of bow tie $= \frac{1}{2} \times 3 \times h + \frac{1}{2} \times 5 \times h'$.

So you have to find h and h' **in terms of** x.

The key is to show the following:

• The two triangles XAB and $XA'B'$ are similar.

Proof: $\angle AXB = \angle A'XB'$ (vertically opposite angles)
The two ends of the bow tie are vertical, and hence parallel:

$\therefore$ $\angle XAB = \angle XA'B'$ (alternate angles)

   $\angle XBA = \angle XB'A'$ (alternate angles).

So triangles XAB and $XA'B'$ are similar. **QED**

$\therefore$ $h : h' = AB : A'B'$

   $= 3 : 5$ (since the triangles XAB and $XA'B'$ are similar)

$h' = x - h$

$\therefore$ $5h = 3h'$

   $= 3(x - h)$

$\therefore$ $h = \frac{3}{8}x, \ h' = \frac{5}{8}x.$

∴ Area of bow tie = $\frac{1}{2}\times3\times\left(\frac{3}{8}x\right)+\frac{1}{2}\times5\times\left(\frac{5}{8}x\right)=\frac{17}{8}x$. **QED**

5 **Claim:** The two shaded rectangles always have equal areas.

Proof: The diagonal of any rectangle $ABCD$ splits the rectangle into two congruent right-angled triangles: ABD and CDB (since $AB = CD$, $BD = DB$ is a common side, and $DA = BC$).

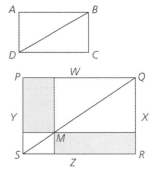

- The diagonal SQ splits the rectangle $PQRS$ into two congruent triangles: $\triangle PSQ$ and $\triangle RQS$.
  ∴ Area $(\triangle PSQ)$ = area $(\triangle RQS)$.

- The diagonal MQ splits the rectangle $WQXM$ into two congruent triangles: $\triangle WMQ$ and $\triangle XQM$.
  ∴ Area $(\triangle WMQ)$ = area $(\triangle XQM)$.

- The diagonal SM splits the rectangle $YMZS$ into two congruent triangles: $\triangle YSM$ and $\triangle ZMS$.
  ∴ Area $(\triangle YSM)$ = area $(\triangle ZMS)$

∴ Area $(PWMY)$ = area $(\triangle PSQ)$ − area $(\triangle WMQ)$ − area $(\triangle YSM)$

$\qquad\qquad$ = area $(\triangle RQS)$ − area $(\triangle XQM)$ − area $(\triangle ZMS)$

$\qquad\qquad$ = area $(MXRZ)$. **QED**

6 **a** Area = a^2 t.u.

Proof: When $a = 1$, the 1 by 1 triangle has area 1 t.u.

The 2 by 2 triangle is obtained from the 1 by 1 triangle by adding a strip of 3 small triangles along the bottom.
∴ Area = $1 + 3 = 4 = 2^2$.

The 3 by 3 triangle is obtained from the 2 by 2 triangle by adding a strip of 5 small triangles along the bottom.
∴ Area = $1 + 3 + 5 = 9 = 3^2$.

The a by a triangle is obtained from the $a-1$ by $a-1$ triangle by adding a strip of $2a-1$ small triangles.
∴ Area = $1 + 3 + 5 + \ldots + (2a-1)$
$\qquad$ = a^2.

(See the solution to question **5c** on page 77 of Book 2). **QED**

b Area = $2ab$ t.u.

Proof: A 1 by 1 parallelogram contains two 1 by 1 triangles;
∴ Area = $2 = 2 \cdot 1 \cdot 1$.

An a by b parallelogram contains a horizontal strips, and each horizontal strip contains b 1 by 1 parallelograms;

$\therefore$ Area = (area of a 1 by 1 parallelogram)
   $\times$ (number of 1 by 1 parallelograms)

$= 2 \times ab$. **QED**

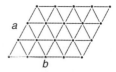

c Area = ab t.u.

Proof: Each a by b triangle is equal to half of an a by b parallelogram.

$\therefore$ Area (a by b triangle) = $\frac{1}{2} \times$ area (a by b parallelogram)

$= \frac{1}{2} \cdot 2ab = ab$. **QED**

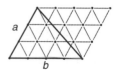

Note: The formula in part **c** helps to explain the answer in part **a** (since the a by a equilateral triangle in part **a** has $a = b$).

d The formulae remain the same!

17 *A sequence of triples*

1 **a** 3, 5, 7, … suggests that the next two entries should be 9, 11. The first entry is 2.1 + 1, the second entry is 2.2 + 1, so you expect that the nth entry will be $2n + 1$.

   b 4, 12, 24, … is not quite so easy. But it looks like: add 8, add 12. This suggests: add 16, add 20, … . If this is correct, then the next two entries will be 40, 60.

   A formula for the nth entry in the second column is more difficult; and it would be rash to generalise algebraically until you have more evidence that 40 and 60 are correct.

   c 5, 13, 25, … suggest the same pattern: add 8, add 12, … . This is supported by the observation that in the first triple $c = 5 = b + 1$; in the second triple $c = 13 = b + 1$; in the third triple $c = 25 = b + 1$. This suggests that the next two entries will be 41, 61.

   If all this is correct, then a formula for the nth entry in the **third** column will follow immediately once you have a formula for the nth entry in the **second** column (you only have to add 1).

2 a $3^2 + 4^2 = 5^2$; $5^2 + 12^2 = 13^2$; $7^2 + 24^2 = 25^2$.

b Does $9^2 + 40^2 = 41^2$? Does $11^2 + 60^2 = 61^2$?

3 The pattern you found in question **2** is mathematically very interesting. This should strengthen your faith in the guesses you made in question **1**. So it is now time to find a formula for the nth entries in the second and third columns.

a Every entry is a multiple of 4. If you take out a factor of 4, and write each entry according to the 'add 8, add 12, add 20, … ' rule you gave in question **1b**, then the sequence of entries in the second column looks like this:

4×1; $4 \times (1 + 2)$; $4 \times (1 + 2 + 3)$; $4 \times (1 + 2 + 3 + 4)$;
$4 \times (1 + 2 + 3 + 4 + 5)$; … .

So the nth entry in the second column is simply four times the sum of the first n positive integers. To find a formula for the nth entry in the second column, you need to know (see Book 2, Challenge 16 question **4b**) that

$$1 + 2 + 3 + 4 + … + n = \tfrac{1}{2}n(n + 1).$$

So the nth entry in the second column is
$4 \times \tfrac{1}{2}n(n + 1) = 2n(n + 1)$.

b The nth entry in the third column is $2n(n + 1) + 1$.

c The nth triple (a, b, c) is $(2n + 1, 2n(n + 1), 2n(n + 1) + 1)$.

$\therefore\ c^2 = [2n(n + 1) + 1]^2$

Use the fact that $(x + 1)^2 = x^2 + 2x + 1$, to write this as

$c^2 = [2n(n + 1)]^2 + 2[2n(n + 1)] + 1$

$\quad = [2n\,(n + 1)]^2 + 4n^2 + 4n + 1$

$\quad = [2n\,(n + 1)]^2 + (2n + 1)^2$

$\quad = b^2 + a^2$. **QED**

4 a $5^2 = 12 + 13$; $7^2 = 24 + 25$; $9^2 = 40 + 41$.

Claim: Each triple (a, b, c) in the sequence satisfies $a^2 = b + c$.

Proof: $(2n + 1)^2 = (2n + 1)\,(2n + 1)$

$\qquad\qquad = 2n\,(2n + 1) + 1 \,.\, (2n + 1)$

$\qquad\qquad = 2n\,(n + 1 + n) + (2n + 1)$

$\qquad\qquad = 2n\,(n + 1) + 2n \,.\, n + (2n + 1)$

$$= 2n(n+1) + [2n(n+1) + 1]$$

$$\therefore \quad a^2 = b + c. \quad \textbf{QED}$$

b Second triple: $5 + 12 + 13 = 30 = 6 \times 5 = 6a$.

Third triple: $7 + 24 + 25 = 56 = 8 \times 7 = 8a$.

Claim: The nth triple (a, b, c) satisfies $a + b + c = 2(n + 1) \times a$.

Proof: $a + b + c = (2n + 1) + 2n(n + 1) + [2n(n + 1) + 1]$

$$= (2n + 1) + (2n^2 + 2n) + [2n^2 + 2n + 1]$$

$$= (2n + 1) + 4n^2 + 4n + 1$$

$$= (2n + 1) + (2n + 1)^2$$

$$= (2n + 1)[1 + (2n + 1)]$$

$$= (2n + 1)[2n + 2]$$

$$= a \times [2(n + 1)]. \quad \textbf{QED}$$

5 The sequence of triples has gaps: that is, missing triples with $a = 4$, $a = 6$, and so on. To find the missing entries in the second column, the single step 'add 8' must be split into two – steadily increasing – steps.
It is tempting to try 'add 3, add 5' – but this does not work when you try to split 'add 12'.

3	4	5
4	?	?
5	12	13
6	?	?
7	24	25
8	?	?

The right split is: add $3\frac{1}{2}$, add $4\frac{1}{2}$, add $5\frac{1}{2}$, … . But why?

You want the new triples to fit into the sequence.

To get a triple with $a = 4$, you need to choose n to satisfy

$$a = 2n + 1 = 4$$

$$\therefore \ n = \tfrac{3}{2}.$$

[This makes sense: $(3, 4, 5)$ is the first triple; $(5, 12, 13)$ is the second triple. And the extra term is term number $1\frac{1}{2}$!]

Then

$$b = 2n(n + 1) = 2 \times \tfrac{3}{2}\left(\tfrac{3}{2} + 1\right) = 3 \times \tfrac{5}{2} = \tfrac{15}{2}; \ c = b + 1 = \tfrac{17}{2}.$$

Finally you can then check that

$$a^2 + b^2 = 4^2 + \left(\tfrac{15}{2}\right)^2 = \left(\tfrac{17}{2}\right)^2.$$

COMMENTS & SOLUTIONS

18 Folding squares, equilateral triangles and regular hexagons

1 You may have met this folding construction as the first step in folding many paper aeroplanes and other shapes. The rectangle you start with can be any rectangle – an A4 sheet of paper will do.

- Fold the short side AB onto AB', so that it lies exactly along AD. Crease along AX.

- Before unfolding, fold along XB'.

Claim: $ABXB'$ is a square.

Proof: $AB = AB'$ (since AB folds exactly on top of AB')

$\quad\quad BX = B'X$ (since BX folds exactly on top of $B'X$)

$\quad\quad \therefore\ ABXB'$ is a kite.

$\quad\quad BX \parallel AB'$ (opposite sides of the starting rectangle)

$\quad\quad \therefore\ ABXB'$ is a rhombus.

$\quad\quad \angle XBA = 90°$ (since $ABCD$ is a rectangle)

$\quad\quad \therefore\ ABXB'$ is a square. **QED**

Finally unfold to obtain the original rectangle with the two fold lines AX and XB'.

2 **Claim:** Triangle RSZ is isosceles with base RS and apex Z.

Proof: You have to explain why $ZR = ZS$.

When you fold along XY, PS folds exactly on top of QR and the point Z stays fixed.

$\therefore\ RZ = SZ$

$\therefore\ $ Triangle RSZ is isosceles. **QED**

3 a If Z is any point on XY, then $RZ = SZ$ (for the same reason as in question **2**). So you have to find the point Z' on XY with $RZ' = RS$. To find Z' fold RS so that R folds 'on top of itself', while S folds onto some point Z' on the line XY. Mark the point Z' exactly beneath S.

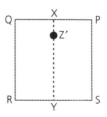

Finally fold along RZ', and along SZ' to mark $\triangle RSZ'$.

b RS folds onto RZ'; $\therefore\ RS = RZ'$.

Z' lies on XY; $\therefore\ RZ' = SZ'$.

$\therefore\ RS = RZ' = SZ'$, so triangle RSZ' is equilateral.

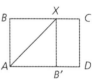

c There is just one triangle congruent to *RSZ'* – namely *RSZ'* itself.

d There are eight triangles congruent to *RYO*: these are *RYO*, *SYO*, *SMO*, *Z'MO*, *Z'NO*, *RNO*, *RNJ*, *SMI*.

e There are four triangles similar to *RSZ'*: these are *RSZ'* itself, *ROJ*, *SOI*, *Z'KL*.

There are twenty triangles similar to *RYO*: *RYO*, *SYO*, *SMO*, *Z'MO*, *Z'NO*, *RNO*, *RNJ*, *SMI*, *Z'XK*, *Z'XL*, *RSI*, *SRJ*, *RQK*, *SPL*, *Z'YR*, *Z'YS*, *SNR*, *RMZ'*, *RMS*, *SNZ'*.

f *RO* is the hypotenuse of triangle *RYO*; and *OM* is the shortest side of triangle *SMO*.

SR folds onto *SZ'* with the crease along *SN*.

∴ *YO* folds onto *MO*.

∴ *OM = OY*.

∴ *RO : OM = RO : OY*.

Triangles *RYO* and *RMS* are both 30°- 60°- 90° right-angled triangles, so must be similar triangles.

∴ *RO : OY = RS : SM*.

M is the midpoint of *SZ'*, and *RS = Z'S*.

∴ *RS : SM = 2 : 1*

∴ *RO : OY = 2 : 1*.

g All angles are equal to 30°, or 60° or 90°.

5 **a, b** In any triangle *ABC*, let *L* be the midpoint of the side *BC*, let *M* be the midpoint of the side *CA*, and let *N* be the midpoint of the side *AB*.

Then the three medians *AL*, *BM*, *CN* all pass through a single point – called the **centroid** of the triangle.

The triangle balances perfectly on a knife edge placed along the median *AL*; similarly for *BM* and for *CN*.

It follows that the point *G* where the three medians meet is the **centre of gravity** of the triangle *ABC*.

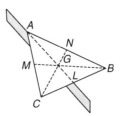

6 Fold Z' so that it lies exactly over O, and crease along AB. Fold R so that it lies exactly over O, and crease along CD. Fold S so that it lies exactly over O, and crease along EF. Then $ABCDEF$, is a regular hexagon.

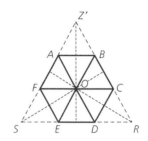

19 Smells, Bells, primes and rhymes

1 a With 1 plate (all three items on one plate) – just 1 way;
with 2 plates (two items on one plate and one on the other) – 3 ways;
with 3 plates (all items on different plates) – 1 way.

b 15 ways.

1 plate: 1 way.

2 plates: (3 + 1) 4 ways; (2 + 2) 3 ways.

3 plates: (2 + 1 + 1) 6 ways.

4 plates: 1 way.

2 d

```
                    1
                1       2
            2       3       5
        5       7      10      15
     15     20      27      37      52
   52     67     87    114    151    203
203   255   322   409   523   674   877
```

e The last entries in each of the first three rows are the first three Bell numbers 1, 2, 5.

The last entry in the fourth row (15) is the fourth Bell number from question **1b**. So it looks as though the last entry in the nth row is the nth Bell number.

If this is correct, then the method certainly provides an easy way to generate the Bell numbers: for example, the fifth, sixth and seventh numbers produced in this way are 52, 203, 877 – and these are in fact the fifth, sixth and seventh Bell numbers.

But it leaves you with a difficult problem: Why should the last number in the nth row of this Bell triangle be equal to the number of different ways in which Alys can serve up n different breakfast items?

3 **a** $30 = 30$, $30 = 15 \times 2$, $30 = 10 \times 3$, $30 = 6 \times 5$, $30 = 2 \times 3 \times 5$.

b 15 ways.

1 factor: $210 = 210$.

2 factors: $210 = 2 \times 105$; $210 = 3 \times 70$; $210 = 5 \times 42$; $210 = 7 \times 30$; $210 = 6 \times 35$; $210 = 10 \times 21$; $210 = 14 \times 15$.

3 factors: $210 = 2 \times 3 \times 35$; $210 = 2 \times 5 \times 21$; $210 = 2 \times 7 \times 15$; $210 = 3 \times 5 \times 14$; $210 = 3 \times 7 \times 10$; $210 = 5 \times 7 \times 6$.

4 factors: $210 = 2 \times 3 \times 5 \times 7$.

c $30 = 2 \times 3 \times 5$ is equal to the product of 3 distinct prime factors. Factorising 30 is exactly like choosing servings with 3 items.

$210 = 2 \times 3 \times 5 \times 7$ has 4 distinct prime factors. Factorising 210 is exactly like choosing servings with 4 items.

4 **a** *abcd*;

aacd, abad, abca, abbd, abcb, abcc; aacc, abab, abba;

aaad, aaca, abaa, abbb;

aaaa.

c The four lines are just like the four items in Alys's breakfast extra (see question **1b**). The rhyming pattern tells you which items are served on the same plate.

Extension

5 The underlined numbers in the Bell triangle on the right appear to be equal to the sum of the numbers in each previous row ($1 = \underline{1}$; $1 + 2 = \underline{3}$; $1 + 3 + 5 = \underline{10}$; etc.). Is this always true?

```
          1
       1     2
    2     3     5
 5     7    10    15
15  20   27   37   52
 •    •    •    •    •    •
```

20 *Folding regular (and irregular) pentagons*

1 This is hard! You must first decide whether to try to prove that the pentagon is regular, or to prove that it is not.

If you want to **prove** that the pentagon obtained in this way from an A4 sheet is regular, you are not allowed to start by **assuming** what you are trying to prove!

However, it may help to change the question slightly and ask:

> What dimensions would the original rectangle *ABCD* have to be if the final pentagon *ARSTU* is to be regular?

If you take this approach, you may clearly assume that the final pentagon *ARSTU* is regular, and then work out what the ratio *AB:AD* of the sides of the original rectangle has to be.

If this ratio *AB:AD* matches that for A4 paper, then fine. If not, then the pentagon *ARSTU* obtained from an A4 sheet cannot be regular.

So one way to get started is to **assume** that the final pentagon is regular, and then work backwards to calculate the dimensions of the original rectangle. This has the advantage that you then know that each corner angle must be 108°, and you can use this to work out what all the other angles on your folded sheet have to be. (At some stage you will have to do a little elementary trigonometry to work out what the ratio *AB:AD* of the sides of the original rectangle would have to be.) The details are given below.

Claim: The pentagon *ARSTU* is **not** regular.

Proof: The plan is to suppose that the final pentagon *ARSTU* is regular, and then to prove that the shape of the original rectangle cannot be A4.

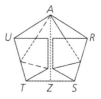

The angle at each corner of the A4 sheet of paper is 90°. The angle at each corner of a regular pentagon is 108°.

∴ ∠*UAY* = 90°, ∠*UAR* = 108°

∴ ∠*YAR* = 108° − 90° = 18°.

Similarly ∠*DAX* = 18°.

∴ ∠*XAY* = 108° − 2 × 18° = 72°

∴ ∠*ZAY* = 36°, and ∠*ZAB* = 36° + 18° = 54°.

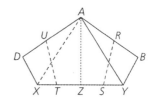

∴ The triangle ABC would have to be a right-angled triangle with angles 54° and 36°.

∴ $\frac{CB}{BA}$ = tan $\angle CAB$ = tan 54° = _____ .

On the other hand, the ratio of the edge lengths of a piece of A4 paper is $\sqrt{2}$; and $\sqrt{2} \neq$ tan 54°.

So if the original rectangle is an A4 sheet, then the pentagon $ARSTU$ is definitely not a regular pentagon. **QED**

2 This is also hard! There are four important things to note.

- The two sides of the original strip of paper are parallel.

- The width of the strip is constant.

- Angles which fold on top of each other are equal.

- If the strip is folded at a point X, then the angles at X add up to 180°.

Claim: The knotted pentagon is a regular pentagon.

Proof: Suppose a strip of constant width d is 'knotted', gently pulled tight and flattened. Then we get a pentagon $ABCDE$ with the following three properties.

1 The line segment AB is parallel to EC (edges of strip of paper); similarly EA is parallel to DB, AC is parallel to ED and DA is parallel to CB.

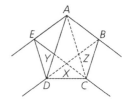

(**Note:** In tying the knot, the strip of paper never runs horizontally; so we simply do not know whether EB is also parallel to DC.)

2 The perpendicular distances from AB to EC, and from EA to DB, and from AC to ED, and from DA to CB are all equal to d, the width of the strip of paper.

3 At each fold, such as the fold over the edge EA (where the strip comes up from DC to EA, and then folds over the back and through to CB), the total angle on the edge of the strip at E (that is, the angle on the front strip, $\angle DEA$, and the angle over the fold, $\angle AEC$) is 180°.

Let X be the point where CE and DB cross. Similarly, let Y be the point where CE and AD cross; and let Z be the point where DB and AC cross.

Then it follows (from **1**) that $XEAB$ is a parallelogram. Using AE as 'base', we have area $(XEAB) = AE.d$ (by **2**);

Using AB as 'base', we have area $(XEAB) = AB.d$ (by **2**).

$\therefore AE = AB$.

Similarly $AYCB$ is a parallelogram, and
area $(AYCB) = AB \cdot d = BC \cdot d$, so $AB = BC$.
And $AZDE$ is a parallelogram, with
area $(AZDE) = AE \cdot d = ED \cdot d$, so $AE = ED$.

It remains to show that $DC = DE$, and that all five angles are equal.

$\angle BAC = \angle BCA$ (since $\triangle BAC$ is isosceles)

$\qquad = \angle CAD$ (since BC is parallel to AD).

Also

$\angle EAD = \angle EDA$ (since $\triangle EAD$ is isosceles)

$\qquad = \angle DAC$ (since DE is parallel to AC).

$\therefore \angle BAC = \angle EAD = \angle CAD$.

So each of these angles is equal to $36°$ (by property **3**, applied to one of the folds of the strip at A).

$\angle BAC = \angle ACE$ (since BA is parallel to CE)

$\qquad = \angle CED$ (since CA is parallel to DA).

Finally, we know that $\triangle BAC$ is congruent to $\triangle EAD$ (since $BA = EA$, $BC = ED$, and $\angle BAC = \angle EAD$, $\angle BCA = \angle EDA$, so $\angle ABC = \angle AED$).

$\therefore AC = AD$

$\therefore \triangle ADC$ is isosceles, so $\angle ADC = \angle ACD$.

$\therefore \angle XDC = \angle XCD$
   (**Proof:** DB is parallel to EA, so $\angle ADB = \angle EAD$;
   we showed above that $\angle EAD = \angle BAC$;
   and $\angle BAC = \angle ACE$, since EC is parallel to AB.)

$\therefore \angle XDC = \angle XCD = 36°$, so $\angle DCE = \angle DEC$, $\triangle DCE$ is isosceles and $DC = DE$.

But then $\angle EAB = \angle EDC = \angle BCD = 3 \times 36° = 108°$;

$\therefore \angle AED + \angle ABC = 3 \times 180° - 3 \times 108°$ (sum of angles in a
$\qquad\qquad\qquad\qquad\qquad\qquad$ pentagon $= 3 \times 180°$)

$\qquad\qquad = 216°$.

Also $\angle AEX = \angle ABX$ (opposite angles of the parallelogram $XEAB$).

Hence $\angle AED = \angle ABC = 108°$.   **QED**

21 *Puzzles*

1   The answer is slightly counter-intuitive.

Let W be the winch.

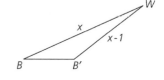

Let B be the first position of the bow of the boat, with the rope WB of length x metres.

Let B' the second position of the bow of the boat after 1 metre of rope has been winched in; so WB' has length $x - 1$ metres.

$\therefore \quad WB' + B'B > WB$

$\therefore \quad x - 1 + B'B > x$

$\therefore \qquad\qquad B'B > 1.$

2   Let the two numbers be x and y.

You have to decide which is bigger: $\frac{x^2 + y^2}{2}$ or $\left(\frac{x+y}{2}\right)^2$.

So you have to decide whether:

the difference $\frac{x^2+y^2}{2} - \left(\frac{x+y}{2}\right)^2$ is positive or negative.

$$\frac{x^2+y^2}{2} - \left(\frac{x+y}{2}\right)^2 = \frac{2(x^2+y^2) - (x^2 + 2xy + y^2)}{4}$$

$$= \tfrac{1}{4}(x^2 + y^2 - 2xy)$$

$$= \tfrac{1}{4}(x - y)^2.$$

Since $(x - y)^2 \geq 0$ always, it follows that

$$\frac{x^2+y^2}{2} - \left(\frac{x+y}{2}\right)^2 \geq 0$$

$$\therefore \qquad\qquad \frac{x^2+y^2}{2} \geq \left(\frac{x+y}{2}\right)^2.$$

3   Suppose first that a square peg $ABCD$ is to be fitted into the circular hole – centre O, radius r.
Triangle AOB is a right-angled isosceles triangle;

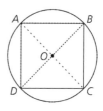

$\therefore \quad AB = r\sqrt{2}.$

$\therefore$ Area $(ABCD)$: area (circle) $= 2r^2 : \pi r^2 = 2 : \pi.$

Suppose next that the hole is square, with side $2r$, and the peg is round (with radius r).

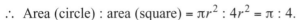

∴ Area (circle) : area (square) $= \pi r^2 : 4r^2 = \pi : 4$.

Now $\pi > 3$, so $\pi^2 > 9$.

∴ $\pi^2 > 8$

∴ $\frac{\pi}{4} > \frac{2}{\pi}$.

So a round peg in a square hole occupies a larger fraction of the available space than does a square peg in a round hole.

4 Write 1D in place of 'the answer to 1 Down', and 4A in place of 'the answer to 4 Across'.

1D is a 3-digit cube. The only 3-digit cubes are:

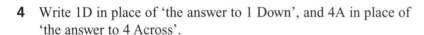

$5^3 = 125$; $6^3 = 216$; $7^3 = 343$; $8^3 = 512$; $9^3 = 729$.

The units digit of this cube is also the units digit of 4A – which is a square. The units digit of any square has to be 0, 1, 4, 5, 6, or 9.

∴ there are just three possibilities for 1D: namely **i** 729; **ii** 216; or **iii** 125.

i If 1D = 729, then 1A would be a 2-digit square between 70 and 79 – which is impossible.

ii If 1D = 216, then 1A = 25; 2D would then be a 2-digit square between 50 and 59, which is impossible.

iii So the only possibility is 1D = 125.

Then 4A = 25 (since there is only one 2-digit square ending in 5),

  1A = 16,

  2D = 64.

The only 3-digit square ending in 24 is $324 = 18^2$.

Moreover, then $32 = 2^3 \times 2^2$ is equal to a cube times a square.

So there is exactly one solution. **QED**

5 Imagine the longest possible ladder.

The ladder has to get round the corner.

∴ At its tightest fit the ends of the ladder will just touch the two walls and the inside corner.

∴ (Maximum length of ladder) = (shortest line through corner).

Because the two corridors have the same width, this line segment occurs at an angle of 45° (it is not easy to prove this).

The ladder and the walls then form an isosceles right-angled triangle with legs of length 6 metres.

So the longest possible ladder is $6\sqrt{2}$ metres.

6 a Let the middle integer be n.

Then the three integers are $n - 1, n, n + 1$, with sum $(n - 1) + n + (n + 1) = 3n$.

So the sum is always divisible by 3.

b With five consecutive integers, let the middle integer be n.

Then the five integers are $n - 2, n - 1, n, n + 1, n + 2$, with sum

$$(n - 2) + (n - 1) + n + (n + 1) + (n + 2) = 5n.$$

So the sum is always divisible by 5.

The same argument shows that 'the sum of k consecutive integers is always a multiple of k provided k is odd'.

c Notice that $1 + 2 + 3 + 4 = 10$, and $2 + 3 + 4 + 5 = 14$ are not multiples of 4.

Claim: The sum of four consecutive integers is never divisible by 4.

Proof: Let the four consecutive integers be $n, n + 1, n + 2, n + 3$.

Then their sum is

$$n + (n + 1) + (n + 2) + (n + 3) = 4n + 6$$

$$= 4(n + 1) + 2$$

$$= \text{(multiple of 4)} + 2.$$

Hence the sum of any four consecutive integers is never a multiple of 4.   **QED**

7 Let the shaded rectangle have sides of lengths a and b.

When turning the shaded rectangle, its diagonal has to fit across the narrowest part of the surrounding 4 by 5 rectangle.

The diagonal of the shaded rectangle has length $\sqrt{(a^2 + b^2)}$.

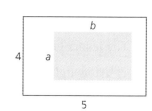

So for a tight fit you can just allow $\sqrt{(a^2 + b^2)} = 4$.

The perimeter of the shaded rectangle has length $2(a + b)$.

You now need an important bit of algebra:

$$(a + b)^2 + (a - b)^2 = (a^2 + 2ab + b^2) + (a^2 - 2ab + b^2)$$
$$= 2(a^2 + b^2).$$

For the shaded rectangle you know that $a^2 + b^2 = 16$.

$\therefore (a + b)^2 + (a - b)^2 = 32$.

This equation tells you that the sum of $(a + b)^2$ and $(a - b)^2$ is always equal to 32; so $(a + b)^2$ can only increase if $(a - b)^2$ decreases.

You want to make the perimeter $2(a + b)$ as large as possible.

So you want $(a + b)^2$ to be as large as possible.

So you want $(a - b)^2$ to be as small as possible: that is, you want $a = b$.

Then $(a + b)^2 = 32$

$\therefore \quad 4\,(a + b)^2 = 128$

$\therefore \quad 2\,(a + b) = 8\sqrt{2}$.

EXTRA 1

Prime numbers

Any prime number $P > 3$ must be odd.

So when you divide P by 6, the remainder must be odd.

$\therefore P = 6k + 1$, or $6k + 3$, or $6k + 5$ for some integer k.

$6k + 3 = 3(2k + 1)$ is only prime when $k = 0$ (and $P = 3$).

$\therefore P = 6k + 1$ or $6k + 5$ for some integer k.

That is, P is either 1 more or 1 less than a multiple of 6.

EXTRA 2

Atoms in the universe

How many different ways are there for the 60 children to line up?

The first place can be filled in 60 different ways; once this place is filled, the second place can be filled in 59 different ways; once these two places are filled, the 3rd place can be filled in 58 ways; and so on.

So the number of different ways of lining up 60 children is

$$60! = 60 \times 59 \times 58 \times 57 \times 56 \times \ldots\ldots \times 3 \times 2 \times 1$$
$$\approx 8.32 \times 10^{81}.$$

The number of atoms in the universe is not known exactly, but the best current estimates indicate that this number is between 10^{76} and 10^{79}.

EXTRA 3

Hoses

A hose of diameter 4 inches has radius 2 inches and cross-sectional area equal to $\pi \times 2^2$ square inches. Two such hoses have total cross-sectional area 8π square inches.

A hose of diameter 6 inches has radius 3 inches, and so has cross sectional area equal to $\pi \times 3^2 = 9\pi$ square inches.

So the single hose should allow them to complete the job more quickly.

EXTRA 4

Tubes

Let the A4 sheet have length l and width w.

i The first cylinder has height w and circular base of circumference l.

Let r denote the radius of the base of the cylinder.

$\therefore\ 2\pi r = l$

$\therefore\ r = \frac{l}{2\pi}$

$\therefore\ V = \pi r^2.w = \pi \left(\frac{l}{2\pi}\right)^2 w = \frac{l^2 w}{4\pi}.$

ii The second cylinder has height l and circular base of circumference w.

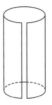

Let r' denote the radius of the base of the cylinder.

$\therefore\ 2\pi r' = w$

$\therefore\ r' = \frac{w}{2\pi}$

$\therefore\ V' = \pi r'^2.l = \pi \left(\frac{w}{2\pi}\right)^2 l = \frac{l w^2}{4\pi}.$

So $V : V' = l : w = \sqrt{2} : 1$ (see Challenge 15 in Book 2).

EXTRA 5

Rhino

Let the baby rhino weight W kg.

Weight depends on volume.

Volume depends on the product of all three dimensions; length, breadth, height.

Notice that $\frac{315}{76} \approx 4$, so the mother rhino is roughly 4 times as long as the baby rhino. Since the baby rhino looks like a rhino, it must be similar in shape to the mother rhino – so the mother rhino must also be roughly 4 times as wide and 4 times as tall.

$\therefore\ \frac{1480}{W} = \left(\frac{315}{76}\right)^3$

$\therefore\ W \approx 20.786.$

Three card trick

Let the three cards be BB (black-black), WW (white-white), BW (black-white).

a There are two ways to put each card down on the table: the first card can go down either B or B; the second card can go down either W or W; the third card can go down either B or W.

∴ There are $2 \times 2 \times 2$ different ways to put all three cards down on the table

(B, W, B), or (B, W, W), or (B, W, B), or (B, W, W),

or (B, W, B), or (B, W, W), or (B, W, B), or (B, W, W).

Four of these possibilities show one black and two whites.

∴ The probability of one black and two whites showing
$= \frac{4}{8} = \frac{1}{2}$.

b There are three black sides altogether.

So when you see a black side, you must be looking at one of these three. Two of the black sides have black on the other side.

∴ The probability that the other side is also black $= \frac{2}{3}$.

c In choosing two sides from the three cards, there are 12 possible pairs. Two of these pairs are WW (you could pick either of the two sides of the all white card and the white side of the black-white card).

Similarly two pairs are BB.

So 8 of the 12 pairs show one black and one white.

∴ The probability of seeing one black and one white
$= \frac{8}{12} = \frac{2}{3}$.

COMMENTS & SOLUTIONS

EXTRA 7

Father Christmas

Father Christmas has two children.

These could be *BB* (boy youngest and boy eldest), *BG*, *GB*, or *GG*.

a   You know that one of the children is a boy.

This rules out *GG* – leaving just three possibilities: *BB*, *BG*, *GB*.

In two of these three possibilities, the younger child is a boy.

∴ The probability that the youngest child is a boy = $\frac{2}{3}$.

b   You know that the older of the two children is a boy.

This rules out *GG* and *BG* – leaving only *BB* and *GB*.

In one of these two cases the youngest child is also a boy.

∴ The probability that the youngest child is a boy = $\frac{1}{2}$.

EXTRA 8

Perfect logic

If Alice could see two green stamps, she would know that her stamp was not green – so she would 'know something'. If she could see two red stamps, she would know that her stamp was not red. So Alice sees neither two reds nor two greens.

Becky knows that Alice cannot see two reds, so if Becky sees a red on Claire's forehead, Becky would know that her own stamp is not red – and so would 'know something'. Therefore Claire knows that Becky cannot see a red on Claire's forehead.

Similarly Becky cannot see a green stamp on Claire's forehead – and Claire knows this.

So Claire concludes that the stamp on her own forehead must be yellow.

Glossary

- This list of symbols, and of common terms and their meanings contains the information you need to make sense of the problems and the given solutions.

- You may need to refer to a mathematics dictionary or a textbook for more detailed explanations.

Symbols

$a > b$	a is greater than b
$a \geq b$	a is greater than or equal to b
$a < b$	a is less than b
$a \leq b$	a is less than or equal to b
$a = b$	a is equal to b
$a \neq b$	a is not equal to b
$a \approx b$	a is approximately equal to b
^-a	*minus a*, or the negative of a
$\sqrt{a}$	the square root of a
a^2	the square of a ($a^2 = a \times a$)
a^n	the nth power of a ($a^n = a \times a \times \ldots \times a$, with n factors equal to a)
$a : b$	the ratio of a to b
$a \equiv b$	a is equivalent to b (or a is identically equal to b)
$\therefore$	therefore
$\Rightarrow$	implies that

Terms

Adjacent   **Adjacent** means *next to*. Two vertices of a triangle, or a polygon, are **adjacent** if they if they are joined by an edge; two edges are **adjacent** if they meet at a vertex.

Alternate angles   If *AB* and *CD* are parallel lines, *X* lies on *AB* and *Y* lies on *CD*, then $\angle AXY$ and $\angle DYX$ are **alternate angles**: $\angle AXY = \angle DYX$.

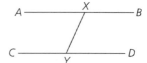

Base   *Integers* are usually written in **base 10**, using digits 0, 1, 2, 3, 4, 5, 6, 7, 8, 9. The right-hand digit of an integer in **base 10** counts units, the next position counts tens, the next counts hundreds and so on. In **base 10**, the number twenty four is written **24**, meaning 2 tens and 4 units.

When numbers are written in **base 3**, only the digits 0, 1, 2 are used. The right-hand digit counts units, the next position counts threes, the next position counts nines (since $9 = 3^2$), and so on. In **base 3**, the number twenty four is written **220**, meaning 2 nines, 2 threes and 0 units.

Billion   Nowadays the word **billion** means a thousand million, or 10^9. (It used to mean a million million – or 10^{12}; it still has this meaning in some European languages, such as German.)

Bisect   To **bisect** something means to cut it into two equal parts. An *angle bisector* is a line which divides a given angle exactly in two. The *perpendicular bisector* of a line segment *AB* is a line through the midpoint of *AB* which is perpendicular to *AB*.

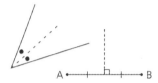

Circumference   The **circumference** of a circle, or of a circular disc, sometimes refers to the geometrical curve which makes up the outer boundary of the circular disc, and sometimes to the length of this curve.

Common   The integers 12 and 15 both have 3 as a factor: 3 is called a **common** factor of 12 and 15. The integer 60 is a multiple of 12; it is also a multiple of 15; 60 is a **common** multiple of 12 and 15. Two adjacent sides of a triangle share a **common** vertex.

Congruent   Two triangles *ABC* and *DEF* are **congruent** if the sides of the two triangles are equal in pairs (*AB = DE, BC = EF, CA = FD*), and the angles are also equal in pairs ($\angle ABC = \angle DEF$, $\angle BCA = \angle EFD$, $\angle CAB = \angle FDE$). If two triangles are **congruent**, they have exactly the same size and shape.
To show that two triangles *ABC* and *DEF* are congruent, it is enough to show
(a) that the three sides are equal in pairs: *AB = DE, BC = EF, CA = FD*. This is called the *SSS congruence criterion;* or
(b) that two corresponding sides and the angle between them are equal: *AB = DE, BC = EF*, $\angle CAB = \angle FDE$. This is called the *SAS congruence criterion*; or
(c) that two corresponding angles and the side between them are equal: $\angle ABC = \angle DEF$, *BC = EF*, $\angle BCA = \angle EFD$. This is called the *ASA congruence criterion*.

Consecutive   Two integers are **consecutive** if they differ by 1, for example, 4 and 5. Two terms of a number sequence, or two events in a sequence of events, are **consecutive** if one comes immediately after the other.

Cube   The word **cube** refers to:
• a solid shape with six square faces, or
• a number like $27 = 3 \times 3 \times 3 = 3^3$, which is equal to the volume of a solid cube with integer side length 3: we say: 27 is the **cube** of 3.

Cuboid	A **cuboid** is a right prism with a rectangular base. A cuboid has three pairs of parallel rectangular faces, any two faces that meet are perpendicular, and three rectangles meet at each of its eight vertices.
Data	**Data** is another name for a list of information of a single kind – usually numerical information; for example, a list of prices of a single item at different times, or the output of different shifts in a single factory.
Decimal fraction	A **decimal fraction** is the decimal of a proper fraction which can be written with denominator a power of 10; for example, $\frac{1}{4} = \frac{25}{100}$, so this gives the *decimal fraction* 0.25.
Decimal place	In any decimal, the position immediately after the decimal point is called the first **decimal place**, that is, the tenths digit. The second position after the decimal point is the second **decimal place**, that is, the hundredths digit.
Denominator	In the fraction $\frac{3}{4}$, the number on the bottom, 4, is the **denominator**. (The number on the top is the *numerator*.)
Diagonal	A **diagonal** of a polygon is any line segment joining two non-adjacent vertices of the polygon.
Difference	The **difference** of two numbers is found by subtracting the smaller number from the larger: for example, the **difference** between $^-2$ and 4 is 6.
Digit	A **digit** is one of the numbers 0, 1, 2, 3, 4, 5, 6, 7, 8, 9. The number 1234 has four **digits**: 3 is the *tens digit* of 1234.
Distribution	A **distribution** is a graph or bar chart which displays the different values of a variable on the x-axis, and which shows the relative frequency with which each value occurs as the corresponding y value.
Divisibility test	A **divisibility test** (or rule) is a quick way of deciding whether a given number is divisible by a fixed number. For example, a number is divisible by 5 precisely when its units digit is 0 or 5. 63 is **divisible by** 9 (since 63 ÷ 9 = 7, with no remainder). 63 is *not* **divisible by** 8 (since 63 ÷ 8 = 7 remainder 7).
Equation	An **equation** is a mathematical sentence in which two numbers or expressions are linked by an '=' sign.
Equilateral	A triangle *ABC* is **equilateral** if its sides *AB*, *BC*, *CA* are all of equal length.
Expression	An **expression** is a collection of mathematical symbols (usually standing for known and unknown numbers, constants and variables), which are added or multiplied together, subtracted or divided. For example, πr^2 is an expression for the area of a circle of radius r, and $2(l + w)$ is an expression for the perimeter of a rectangle of length l and width w.
Factor	9 is a **factor** of 63, because 63 = 9 × 7; 9 is *not* a **factor** of 64.

Formula	A **formula** is an equation which expresses the possible values of one variable in terms of one or more other variables. For example, $C = 2\pi r$ expresses the circumference of a circle of radius r in terms of r; $V = \pi r^2 h$ expresses the volume V of a cylinder in terms of the radius r of the base and the height h of the cylinder.
Fraction	A **fraction** is the number obtained when one integer quantity (the *dividend*, or *numerator*) is divided by a non-zero integer quantity (the *divisor*, or *denominator*).
Frequency	The **frequency** of an event is the number of times the event occurs.
Frequency distribution	A **frequency distribution** is a graph, or bar chart, which displays the different values of a variable on the x-axis, and which shows the *frequency* (or the *relative frequency*) with which each value occurs as the corresponding y value.
Glossary	A **glossary** is a list of specialist words, together with their meanings.
Highest common factor	The **highest common factor** (*hcf*) of two given integers is the largest integer which is a factor of both of the given integers. For example, the *hcf* of 26 and 91 is 13: we write this as *hcf* (26, 91) = 13.
Hypotenuse	The **hypotenuse** of a right-angled triangle is the side opposite the right angle.
Index	When a number, such as 10^4 or 3^4, is written as a power, the power '4' is called the **index** (or exponent).
Integer	An **integer** is any whole number, whether positive, negative or zero; for example, ⁻12, 0, 5, 746.
Isosceles	A triangle ABC is **isosceles** with apex A if $AB = AC$; the side BC opposite the apex A is called the *base* of the isosceles triangle.
Least common multiple	The **least common multiple** (*lcm*) of two given integers is the smallest integer which is an exact multiple of the two given integers; for example, the *lcm* of 26 and 91 is 182.
Linear equation	A **linear equation** is an equation in which all variables appear to the first power; for example, $2x + 3y = 7$. If the formula relating two variables is a linear equation, we say that the relationship between the two variables is **linear**.
Lowest common denominator	The **lowest common denominator** of two given fractions is the *lcm* of the two denominators. For example, the **lowest common denominator** of the fractions $\frac{5}{26}$ and $\frac{17}{91}$ is 182: both fractions can be written with this denominator, $\frac{5}{26} = \frac{35}{182}$ and $\frac{17}{91} = \frac{34}{182}$, and this is the *smallest* such denominator.
Multiple	12 is a **multiple** of 3, since 3 is a *factor* of 12 (12 = 3 × 4). ⁻18 is also a **multiple** of 3 (since ⁻18 = 3 × ⁻6). 13 is *not* a multiple of 3.

Non-terminating decimal	A **non-terminating decimal** is a decimal which goes on for ever – like 0.123 123 123 123 123 123 … , or 1.234 567 891 011 121 314 151 617 181 920 2… . The first of these examples has a *recurring* block: '123', the second example has no recurring block.
Numerator	In the fraction $\frac{3}{4}$, the number on the top, 3, is the **numerator**. (The number on the bottom is the *denominator*.)
Parallel	Two (infinite) straight lines in the plane are **parallel** if they never meet.
Parallelogram	A **parallelogram** is a quadrilateral $ABCD$ in which the side AB is parallel to DC, and the side BC is parallel to AD.
Percentage	To refer to a specific fraction of a given quantity – such as 'one quarter of a rectangle', or 'two thirds of a class of 30 pupils' – we can express this fraction as a **percentage**. To calculate the percentage corresponding to a given fraction, first write the fraction with denominator 100; then take the numerator. For example $\frac{1}{4} = \frac{25}{100}$, so 'one quarter of a rectangle' is the same as '25% of the rectangle'.
Perfect square	An integer is a *square*, or a **perfect square**, if it is equal to the product of some integer by itself; for example $0 = 0 \times 0$, $1 = 1 \times 1$, $4 = 2 \times 2$. So an integer is a **perfect square** if it is equal to the area of a geometrical square with integer length sides.
Perpendicular	Two straight lines are **perpendicular** if they are at right angles to each other.
Plane	A **plane** is a flat surface – that is, a surface in which the straight line joining any two of its points lies wholly in the surface.
Polygon	A **polygon** with three vertices (and three sides) is a triangle; a **polygon** with four vertices is a *quadrilateral*; a **polygon** with five vertices is called a pentagon, one with six vertices is a hexagon, one with eight vertices is an octagon, and one with ten vertices is a decagon. A **polygon** $ABCDE … M$ is an ordered sequence of points $A, B, C, D, E, … , M$ in the plane (called *vertices*) and line segments $AB, BC, CD, DE, … , LM, MA$ (called the sides, or edges), such that any two consecutive edges (such as AB and BC, or MA and AB) have exactly one vertex in common, and any two non-consecutive edges have no points in common at all.
Power	A **power of 10** is a number of the form 10^2 (= 100), or 10^3 (= 1000), or 10^4 (= 10 000). 10^1 (= 10), and 10^0 (= 1) are also **powers of 10**. Similarly, a **power of 2** is a number like 2^0, or 2^1, or 2^2, or 2^3, … .
Prime number	A **prime number** is a positive integer which has exactly two factors – namely itself and 1. 1 is *not* a prime number.

Prism	A **prism** is a three-dimensional solid with polygonal base B, such that any cross-section parallel to the base B is congruent to B. For example, a cuboid is a prism with a rectangular base.
Product	Given a collection of numbers, their **product** is the number obtained when you multiply them all together. For example, the product of 2, 3, 4 is $2 \times 3 \times 4 = 24$.
Proof	Given any precise logical statement, a **proof** of that statement is a sequence of logically correct steps which shows that the statement is true.
QED	**QED** is often written to mark the end of a proof. Q, E, and D are the initial letters of the latin clause *Quod Erat Demonstrandum* – meaning 'which is what was to be proved'.
Quadrilateral	A **quadrilateral** is a polygon with four sides.
Random	When we call an event **random** it has to be the result of a process which can be repeated over and over again (such as tossing a coin, or choosing a number from a given collection of numbers). The word **random** then describes not just one particular event or outcome, but rather the unpredictable nature of the underlying process. The process is **random** if, when we use it over and over again to generate a sequence of outcomes, this sequence displays no regularity of any kind. So no matter how much you know about the outcomes which have occurred up to some point, you cannot use this information to predict individual future events in the sequence.
Ratio	The **ratio** of two quantities is given by a pair of numbers separated by a colon which indicates the relative size of the two quantities. For example, the number of white squares and black squares in the diagram are in the **ratio** 3 : 2. □ ■ ■ □ □
Reciprocal	The **reciprocal** of x is $\frac{1}{x}$; so the reciprocal of $\frac{2}{3}$ is $\frac{3}{2}$.
Recurring decimal	A **recurring decimal** is a decimal like that for $\frac{1}{3} = 0.333\,333\ldots$; or $\frac{1}{6} = 0.666\,666\ldots$; or $\frac{1}{7} = 0.142\,857\,142\,857\ldots$; or $\frac{1}{11} = 0.090\,909\,090\ldots$, which goes on for ever, repeating the same string of digits over and over.
Reflection symmetry	A shape has **reflection symmetry** with mirror line m if, when you swap the two parts on either side of the (dotted) mirror line, the whole shape lands up exactly on top of itself.
Regular polygon	A **regular polygon** is a polygon in which all the sides are equal and all the internal angles are equal. Thus an equilateral triangle is a regular 3-gon, and a square is a regular 4-gon.
Relative frequency	If, in a sequence of N trials, a particular outcome occurs p times – that is, the particular outcome has frequency p – then its **relative frequency** is $\frac{p}{N}$. A **relative frequency chart** displays the overall results of such a sequence of

trials by plotting the possible outcomes, or values of a variable, on the x-axis, and the relative frequency with which each event, or value, occurs as the corresponding y value.

Remainder 10 goes into 23 twice, with **remainder** 3.

Rhombus A **rhombus** is a *quadrilateral* with all four sides of equal length.

Rotational symmetry A rectangle has **rotational symmetry** about its centre P (because it lands up exactly on top of itself when you rotate it through 180° about the point P).

A shape has **rotational symmetry** (or point symmetry) about a point P if it is possible to map it exactly on top of itself by rotating it through some angle less than 360° about P.

Rounding A decimal such as 512.683 can be **rounded to** 1 decimal place as 512.7, or **rounded to** 2 significant figures as 510.

Scattergram Two variables x and y may be loosely related even when they are not related by a simple formula: for example, the value of y may be determined not only by the value of x, but also by some additional random effects. In such cases you can often see the connection between the two variables by plotting known values of the pairs (x, y) in the plane; such a plot is called a **scattergram**.

Segment The line **segment** AB is that part of the infinite line through the two points A and B which lies between A and B (together with the two end points A and B).

Similar Two shapes are **similar** if they can be labelled so that each segment of one shape is k times as long as the corresponding segment in the other shape, for some fixed k. The two most important consequences of this definition are:
  (a) two shapes are **similar** if they have exactly the same shape, but may be of different sizes;
  (b) two triangles ABC, $A'B'C'$ are **similar** precisely when corresponding angles are equal in pairs: $\angle ABC = \angle A'B'C'$, $\angle BCA = \angle B'C'A'$, $\angle CAB = \angle C'A'B'$.

Square An integer is a **square** (sometimes called a *perfect square*) if it is equal to the product of some integer by itself; for example, $0 = 0 \times 0$, $1 = 1 \times 1$, $4 = 2 \times 2$. So an integer is a **square** if it is equal to the area of a geometrical square with integer length sides.

Square root $9 = 3 \times 3$ is the *square* of 3; 3 is the **square root** of $9 = 3 \times 3$. We use the symbol $\sqrt{}$, and write $3 = \sqrt{9}$.

Standard form A number which is written as a number between 1 and 10 multiplied by a power of 10 is in **standard form**. For example, $525 = 5.25 \times 10^2$.

Sum The **sum** of 7 and 16 is $23 = 7 + 16$.

Symmetrical A plane shape is **symmetrical** if it has either rotational or reflection symmetry.

Tally A **tally** keeps count by grouping 'tally marks' in fives: ⊬⊬ ⊬⊬ |||.

Term Given a sequence such as
$$1^2 = 1, 2^2 = 4, 3^2 = 9, 4^2 = 16, \ldots, n^2, \ldots,$$
1 is the first **term**, 4 is the second **term**, and n^2 is the nth **term**.

Terminating decimal A **terminating decimal** is a decimal which terminates, or stops.

Tesselates A plane shape **tesselates** the plane if infinitely many copies of the shape can fit together to cover the whole plane, without overlapping and without leaving any gaps.

Trapezium A **trapezium** is a quadrilateral with at least one pair of parallel edges.

Triangular number A **triangular number** is the number of dots in a triangular array like the one shown here. Thus each **triangular number** is given by a sum of the form $1 + 2 + 3 + 4 + \ldots + n$ for some n.

Vertex A **vertex** of a polygon is a point where two adjacent edges meet. A **vertex** of a three-dimensional solid is a point where three or more edges meet.

Vertically opposite angles If AB and CD are lines which cross at the point X, then $\angle AXC$ and $\angle BXD$ are **vertically opposite angles**: $\angle AXC = \angle BXD$.